HISTORICAL ATLAS OF ARKANSAS

Historical Atlas of Arkansas

Gerald T. Hanson

Carl H. Moneyhon

University of Oklahoma Press : Norman and London

BY GERALD T. HANSON

Arkansas Geography: The Physical Landscape and the Historical-Cultural Setting (Little Rock: 1981)
Historical Atlas of Arkansas (with Carl H. Moneyhon) (Norman, 1989)

BY CARL H. MONEYHON

Republicanism in Reconstruction Texas (Austin, 1980)
Documentary History of Arkansas (Fayetteville, 1984)
Portraits of Conflict: A Photographic History of the Civil War in Arkansas (with Bobby Roberts) (Fayetteville, 1987)
Historical Atlas of Arkansas (with Gerald T. Hanson) (Norman, 1989)

Library of Congress Cataloging-in-Publication Data

Hanson, Gerald T.
Historical atlas of Arkansas.

Includes index.
Bibliography: p.
1. Arkansas—Historical geography—Maps.
I. Moneyhon, Carl H. II. Title.
G1356.S1H3 1989 911′.767 88-40545
ISBN 0-8061-1844-X

The paper in this book meets the guidelines for permanence and durability of the Committee on Production Guidelines for Book Longevity of the Council on Library Resources.

CONTENTS

PREFACE

Any examination of the history of Arkansas will indicate that the area's geography has played a major role in the lives of the human beings who have settled here. The soil and resources of what is now Arkansas have channeled and limited Arkansans' efforts to make a livelihood and have contributed to differences among them. Natural highways and barriers have determined the locations of their settlements and also their economic opportunities and activities. This atlas is intended to draw attention to the complexity of the state's geography and to show how significantly geography has affected the state's history.

If there is a theme to the narratives that accompany the maps in the atlas, it is how the people of Arkansas have attempted to transcend the limits of geography. In earlier days their efforts met with only limited success, but through time new technologies have made possible a degree of freedom from the strict restraints imposed by nature on society here. In particular, breakthroughs in transportation such as the steamboat, the railroad, and the automobile helped create new eras in the state's history. The organization of the maps, especially those following the discussion of land subdivision and boundaries in Part Four, reflects the authors' interpretation of this relationship of geography, people, and technology.

We hope that we have produced an atlas that will provide some new insights into the history of Arkansas for specialists and professional scholars. While they will find little new information here, we hope that the relationships drawn between many different factors will be at least provocative. We also hope that this work will appeal to those who are not professionals, and especially that it will give students in secondary schools and colleges a better understanding of Arkansas.

The authors wish to thank several agencies and individuals who have provided information and advice for this atlas. Among the agencies that provided assistance were the Arkansas History Commission, the University of Arkansas at Little Rock Special Collections Library, the Office of the Arkansas Secretary of State, and the Arkansas Industrial Commission. Special thanks are extended to Russell Baker in the reading room of the History Commission and Bobby Roberts and his staff in Special Collections, University of Arkansas at Little Rock. A particular debt is owed to John Ferguson, the State Historian, for his advice and for taking the time to read through the manuscript and make numerous suggestions. Hester Davis, State Archaeologist, examined the material on Indians and helped us to avoid numerous errors. We also want to thank Jim Anderson for his additional work on the maps. While the comments of readers for a press are seldom welcomed by writers, those who read for the University of Oklahoma Press provided many helpful suggestions. We would also like to thank John N. Drayton of the University of Oklahoma Press and his staff for their encouragement and help. A special thanks is owed to Sarah I. Morrison, who served as the press's editor for this project.

GERALD T. HANSON
CARL H. MONEYHON

Little Rock, Arkansas

PART ONE
PHYSICAL ENVIRONMENT

1. ARKANSAS'S LOCATION WITHIN THE CONTERMINOUS UNITED STATES

1. ARKANSAS'S LOCATION WITHIN THE CONTERMINOUS UNITED STATES

Arkansas is located in the south-central United States. The state extends between 33° and 36° 30′ north latitude and from 90° to 94° 30′ west longitude. Its central position within the lower forty-eight states gives Arkansas high accessibility to national markets and to population centers on both coasts. Its central position within the nation's Sun Belt, those states stretching across the southern United States from California to Virginia, has proven economically important and has helped Arkansas to attract industries and people moving into the South from the Northeast.

Arkansas has common borders with six other states: Texas, Louisiana, Mississippi, Tennessee, Missouri, and Oklahoma. The state ranks twenty-sixth in area among the fifty states, with 53,225 square miles. The estimated population of Arkansas in 1987 was 2,397,000 people, which made it the thirty-third most populous state in the Union.

Arkansas is shaped roughly like a square. It is 245 miles from north to south at its maximum extent and 275 miles from east to west at its widest point. From the northeast corner to the southwest corner of the state the distance is 332 miles.

Arkansas is located along major transitional lines in physiographic and ecological zones of the United States. Physiographically, the state is almost evenly divided between the Interior Plateau, which comprises the Ozark and Ouachita upland subregions, and the Gulf Coastal Plain, which includes the Mississippi River Alluvial Plain and West Gulf Coastal Plain. Ecologically, the state contains both the subtropical coniferous forests typical of the South and the temperate deciduous forests of the northern United States.

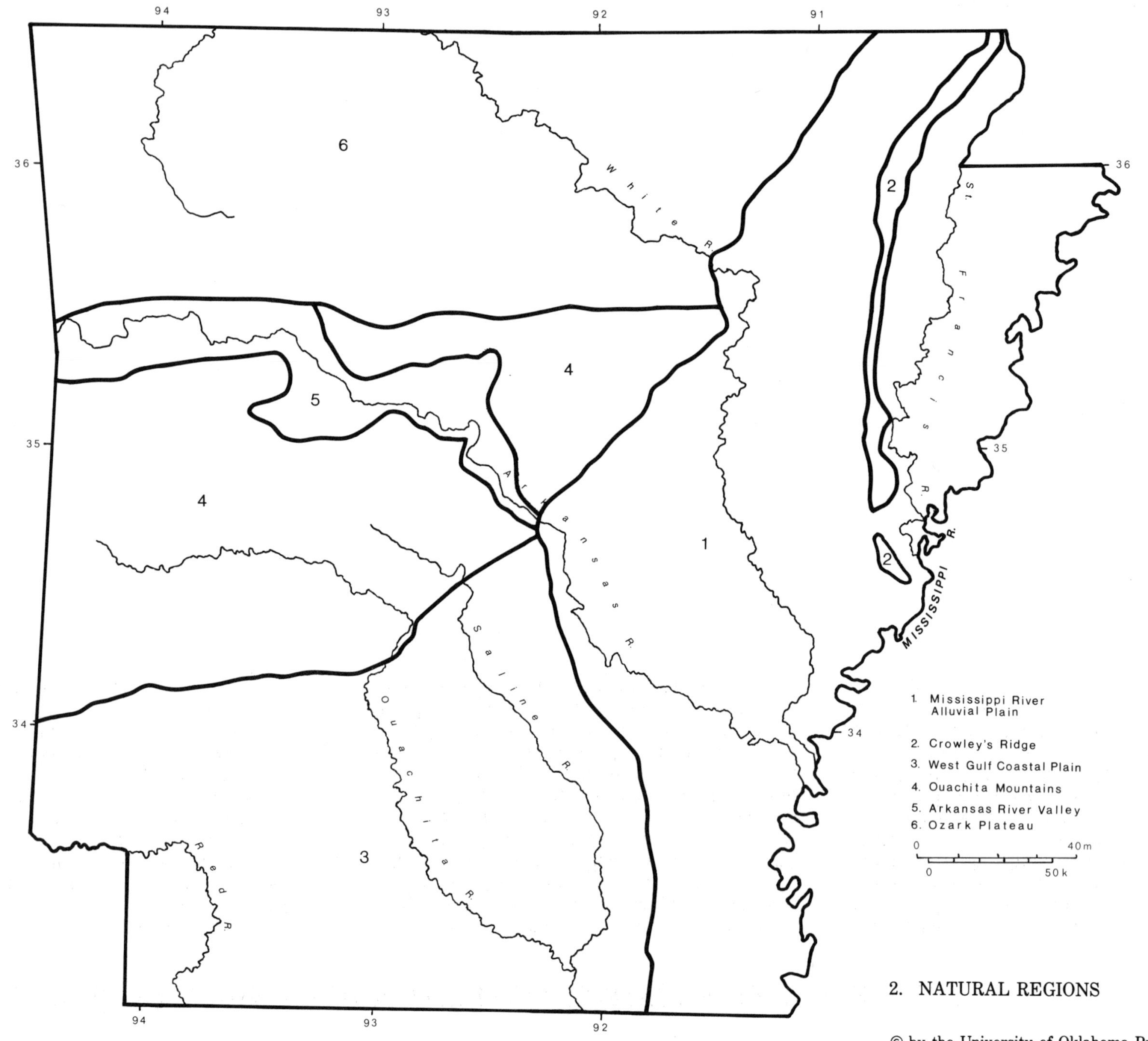

2. NATURAL REGIONS

2. NATURAL REGIONS

A natural region is defined by its topography, climate, soil, vegetation, and water features. The six natural regions of Arkansas offer a composite picture of the state's physical environment.

In the eastern part of the state is the Mississippi River Alluvial Plain (1). The soil of this flat region is several feet deep. The rich alluvium, deposited over the past two million years by the Mississippi River system, provides the base for the modern agriculture extensively practiced here. Before 1850 most of the area was covered by a bottomland hardwood forest, which has now dwindled in area but can be seen along many of the region's rivers. An extensive natural grassland once dominated the center of the Alluvial Plain in Arkansas, Prairie, and Monroe counties. It was long ago converted to rice, soybean, and wheat farms. Surface water is abundant on the Alluvial Plain, where numerous bayous and rivers flow southward.

The natural region of Crowley's Ridge (2) is contained within the Alluvial Plain. The Ridge extends for more than two hundred miles, from southern Missouri to Helena, Arkansas. The area was once covered by a thick layer of wind-blown silt, called loess, which through the last several hundred years has been eroded away by water and wind so that only Crowley's Ridge remains. The soil base is of moderate fertility, but it is easily eroded away. The Ridge, much dissected by running water, has a rugged appearance. Within its upland hardwood forest are eastern hardwoods similar to those of the Appalachian Mountains.

The West Gulf Coastal Plain (3) occupies the southwestern quarter of the state. Its gently undulating topography is broken by many streams flowing from northwest to southeast, and the plain consists of eroded deposits left by running water from the Ouachita River system. The infertile, red-yellow coastal soil is highly leached and offers little potential for agriculture. Only in the river bottoms are the soils fertile enough for crop and animal raising. Coniferous forests dominate the region, and forestry operations comprise its chief economic activity. In addition to the numerous streams, several man-made reservoir lakes are located in this region.

The Ouachita Mountains (4) lie in west-central Arkansas between the Arkansas River valley and the West Gulf Coastal Plain. Within the Ouachita region are the Fourche Mountains, the Novaculite Uplift, and the Athens Piedmont Plateau. The sedimentary rocks in the Ouachitas have been pushed into narrow, tight folds; the ridges and valleys run east and west, similar to a corrugated washboard. The region's soils are thin, infertile, and grey-brown to red-yellow in color. The vegetation is largely coniferous forest; short-leaf pine is native to the area, and loblolly pine has been planted by man. Streams in the region generally flow eastward into the Arkansas or the Ouachita river system. Man-made reservoir lakes scattered across the area include Lake Ouachita, Lake De Gray, Blue Mountain Lake, and Nimrod Lake.

The Arkansas River cuts through the folded-ridge and valley area between the Ozark Plateau and the Ouachita Mountains to form the Arkansas River valley (5), a region that has been greatly modified by erosion caused by running water. The major physical feature here is the Arkansas River, which has cut a wide valley through the region. Soils are richer in the valley than on the Ozark Plateau, since alluvium has been deposited by the river over thousands of years. These deep, fertile deposits have encouraged intensive agricultural use. Vegetation is mixed: the bottomland hardwood forest dominates the valley floors, and coniferous forests cover the linear ridges, on which some temperate deciduous forests are interspersed. Surface water is more abundant here than in the Ozarks; in addition to the Arkansas River, the Petit Jean and Maumelle rivers cross the area.

The Ozark Plateau (6) is a domal mountain structure that has been uplifted and worn down many times over millions of years. The entire surface has been eroded to a plateau surface with more or less uniform elevations among the three subregions in the area. The southernmost unit is the Boston Mountains, which have been tilted slightly so that their southern margin is higher than their northern edge. They are the most rugged of the Ozarks and have been deeply dissected by running water. The Springfield Plateau covers much of northern Arkansas. Here the broad interfluves are at similar elevations, presenting a flat horizon to viewers. Over many thousands of years the streams of the region have carved deeply dissected valleys between the interfluves. East of the Springfield Plateau is the Salem Plateau, which extends into Missouri and is slightly lower than the Springfield Plateau.

Soils on the Ozark Plateau are thin and of moderate fertility. Developed from the limestone and sandstone rocks, they support broadleaf deciduous forests. This is a climax vegetation in this ecosystem, within which coniferous forests are interspersed. Surface water is limited here because the sedimentary rocks are porous, although several large man-made lakes exist in the region.

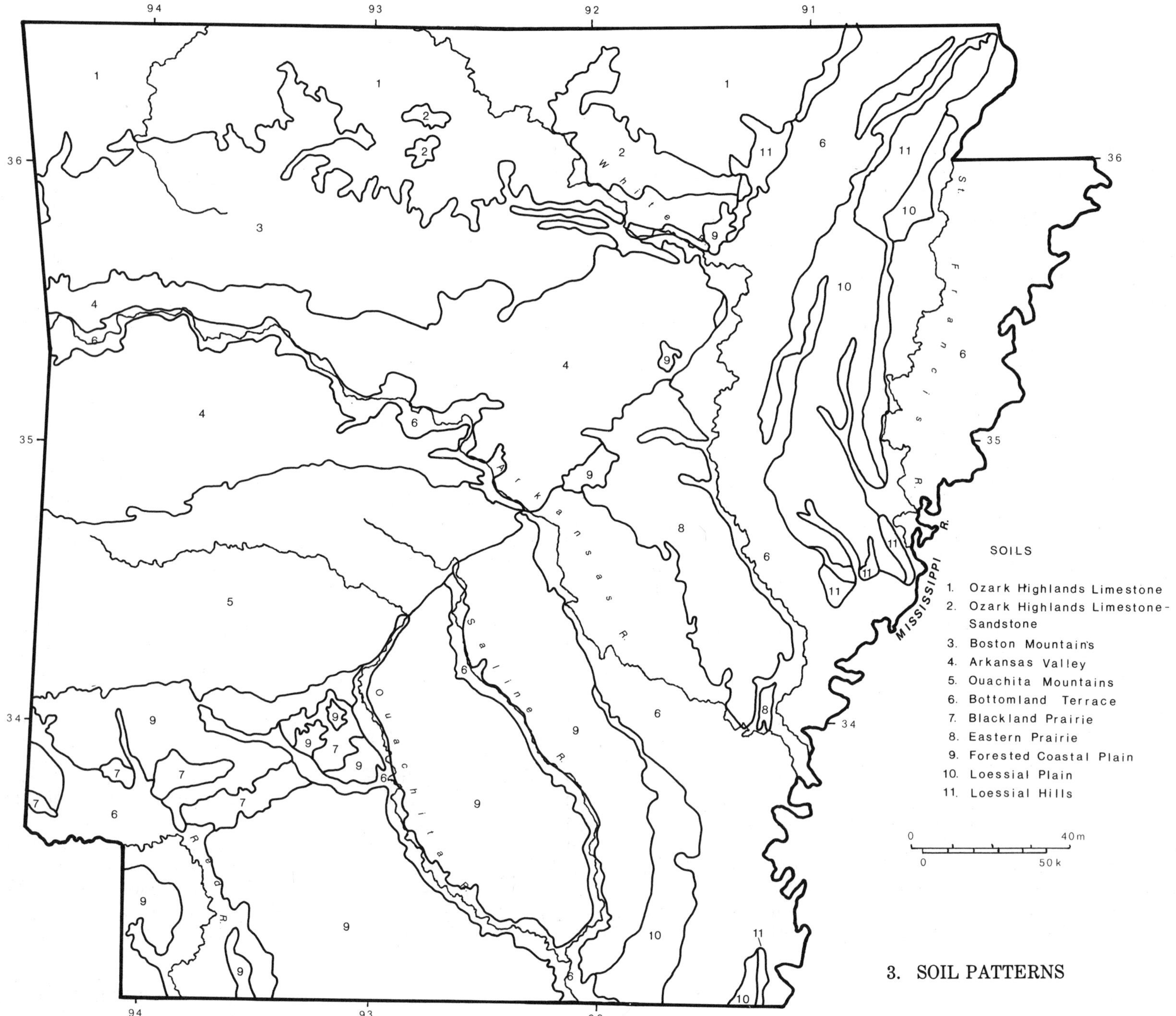

3. SOIL PATTERNS

3. SOIL PATTERNS

Soil groups in Arkansas exhibit considerable variation, reflecting different topographic, geologic, and vegetation conditions. Eleven major groups are identified, each unique with respect to its formation. The general uses of the soils are also important to note.

Three soil groups are found on the Ozark Plateau. These are Ozark Highlands Limestone (1), Ozark Highlands Limestone-Sandstone (2), and Boston Mountains soils (3). Geology obviously played a critical role in the formation of these soils with limestone and sandstone bases. Although generally thin and only moderately fertile, they support a vast hardwood deciduous forest which provides valuable organic material each year. Soils are deeper in the region's narrow valleys, but offer limited agricultural potential, except in Benton and northern Washington counties. Where forests have been removed, general farming is practiced and land is used for pastures, but hardwoods do better than crops in these soils.

Situated in between the Ozark and Ouachita mountains are the Arkansas Valley soils (4). Derived from sandstone and shale rocks, these soils are common on the narrow ridges and valley floors. They are thicker and more fertile than the Ozark soils. Sandy and silty in texture, they support a mixed deciduous and evergreen forest. The valley floors are prime farmland areas where the state's chief crops flourish.

The Ouachita Mountain soils (5) are similar in texture and parent rock material to the Arkansas Valley group, but they are thinner and less fertile. A dense needle-leaf evergreen forest covers the region, and the needles impart considerable acidity to the soils, further reducing their fertility. Coniferous forests are well adapted to these soil conditions.

Much of eastern Arkansas, as well as the major river-valley floodplains, contains bottomland Terrace soils (6) made up largely of alluvium washed into the area by streams. The soils consist of fine clay and silt particles, and they are among the richest in the state, supporting much of Arkansas's commercial agriculture.

A small area in southwestern Arkansas contains the Blackland Prairie (7). Formed from chalk and calcareous marls, the soils here have an upper layer of rich, dark humus, under which lies a clay subsoil. Prairie vegetation dominated these soils historically, but much has been removed for farming.

Eastern Prairie soils (8) consist of a silt-loam upper layer with a compact clay subsoil. These Grand Prairie soils are limited chiefly to Arkansas and Prairie counties. The soil is rich and sustains most of the state's principal crops.

Forested Coastal Plain (9) covers much of southwestern and south-central Arkansas. The gently rolling terrain is covered by a sandy and silty soil formed by the erosion of the Ouachitas. The southern-pine forest adds acidity to the soil. The coastal soils are reddish-yellow in color and often are called podzols. Only in the river valleys of the Red, Ouachita, and Saline rivers does one find a different soil group in this area.

The soils of the Loessial Plains (10) and Loessial Hills (11) are wind-blown silt deposits, with the Loessial Hills soils located primarily on Crowley's Ridge and the Loessial Plains soils on the plains west of the ridge. While the ridge deposits are eroded easily and therefore not congenial for row-crop agriculture, the soils are rich, and on the plains they are used to grow commercial crops.

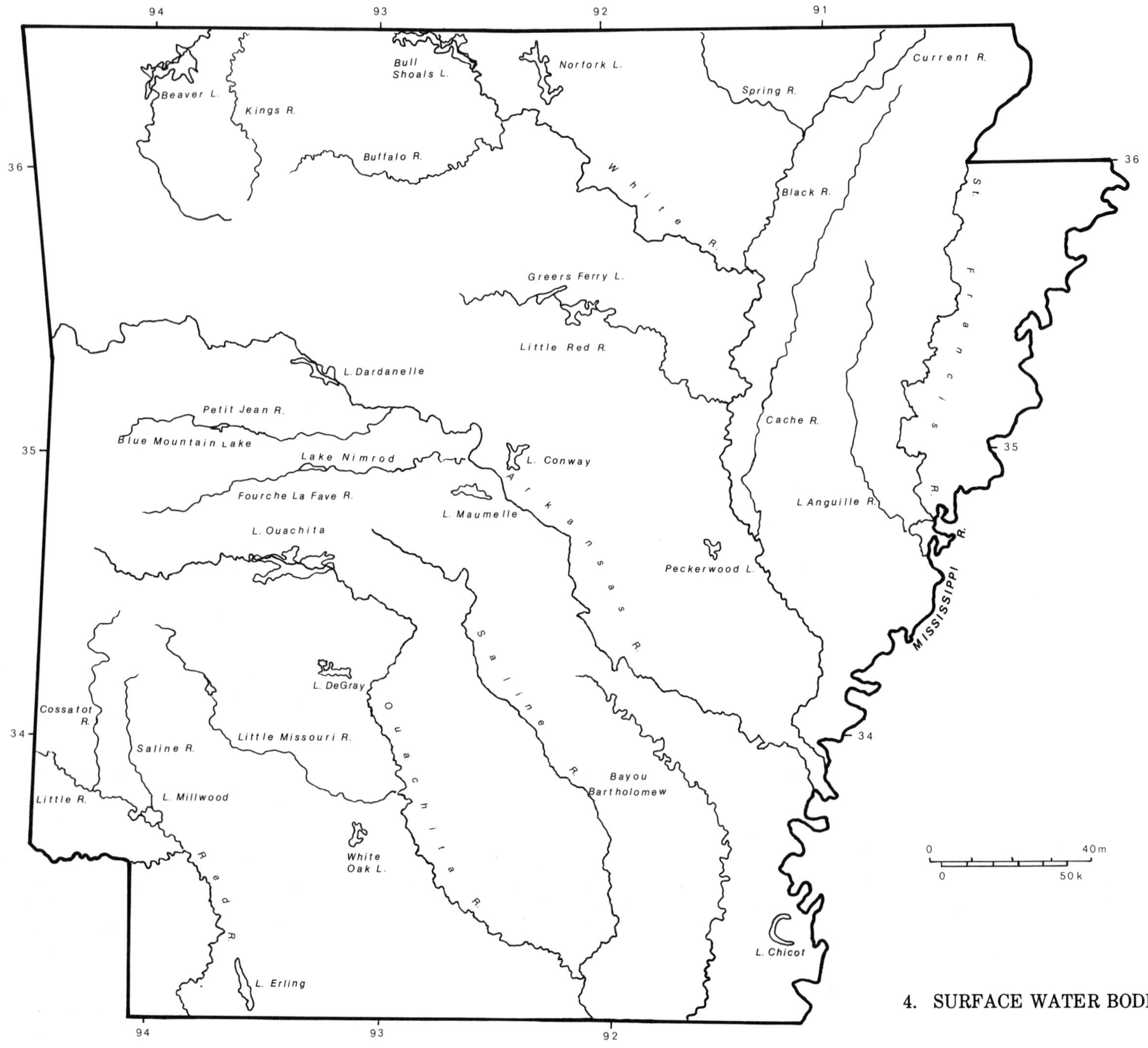

4. SURFACE WATER BODIES

4. SURFACE WATER BODIES

Arkansas has a tremendous abundance of water, both below the surface and above the ground. Numerous streams drain the Mississippi River Alluvial Plain and the West Gulf Coastal Plain, and throughout eastern Arkansas huge acquifers provide subsurface water. Six major river systems drain the state. These rivers have been important historically as transportation routes. Natural lakes are dispersed widely across lowland Arkansas, and in the highlands there are numerous man-made lakes. All these lakes provide considerable recreation for Arkansans and visitors from other states.

The Mississippi River, which constitutes much of the eastern boundary of the state, is Arkansas's largest stream. All of the state's other major river systems drain into it. As a transportation route it brought many settlers into the state.

The Arkansas is the next largest river in the state. Beginning in the Rocky Mountains, about seventy miles northwest of Colorado Springs, the river flows some 1,450 miles to the Mississippi. Entering the state at Fort Smith on the western border, it meanders southeastward across Arkansas. Within the state, the Petit Jean and the Fourche La Fave are its chief tributaries.

The White River rises in the western Boston Mountains in northwestern Arkansas. Its two branches, the West and the Middle Fork, run to the north, joining together just east of Fayetteville. From this junction, the White runs north into Missouri, then reenters the state in northeastern Arkansas. Like many streams that originate in the state's highland regions, the White begins its course as a narrow, rapidly flowing stream, but after leaving the Ozark Plateau it meanders lazily southward across the Alluvial Plain. Major rivers that feed the White are the Buffalo, the Little Red, the Cache, and the Black. The latter is fed by the Spring and Current rivers.

Originating in Missouri, the Saint Francis River enters Arkansas and flows south across the Alluvial Plain, cutting through the remaining bottomland hardwood forest of eastern Arkansas. It is the Saint Francis River that drains the swampy areas of northeastern Arkansas created by the New Madrid earthquake of 1811 and 1812. It is joined near the end of its course by the L'Anguille.

The Ouachita and Saline rivers rise in the Ouachita Mountains. Like typical mountain rivers, they begin as fast-moving streams, but upon entering the Gulf Coastal Plain they turn into meandering waterways. The Saline discharges into the Ouachita in southern Bradley County. The Ouachita continues southward into Louisiana, where it empties into the Mississippi. The Ouachita is also fed by the Little Missouri, on the southwest, and the Bayou Bartholomew, which runs out of southeastern Arkansas and joins the Ouachita in Louisiana.

In southwestern Arkansas the Red River provides a portion of the boundary between Texas and Arkansas. A slow-moving stream, the Red cuts across the alluvial floodplain of Miller County before entering Louisiana. Arkansas tributaries include the smaller Saline and the Little River, with the Little River supported by the Cossatot.

Arkansas has numerous lakes, both natural and man-made. The Ozark and Ouachita regions, for example, have few natural lakes but many man-made ones. The best known of these reservoirs are Beaver, Norfork, Bull Shoals, White Oak, Peckerwood, and Greers Ferry on the Ozark Plateau. In the Ouachitas, dams have been built to create Lakes De Gray, Millwood, Nimrod, and Ouachita. In the Arkansas River valley streams have been dammed to create Lakes Conway, Dardanelle, Maumelle.

The lowland regions contain the greatest amount of surface water, although much of it is in the form of wetlands. Lake Chicot is the largest natural lake in the state. Scattered across the Alluvial Plain are numerous small lakes created by the slow-moving bayous that dominate the area. Lake Erling, in Lafayette County, is typical of these reservoirs.

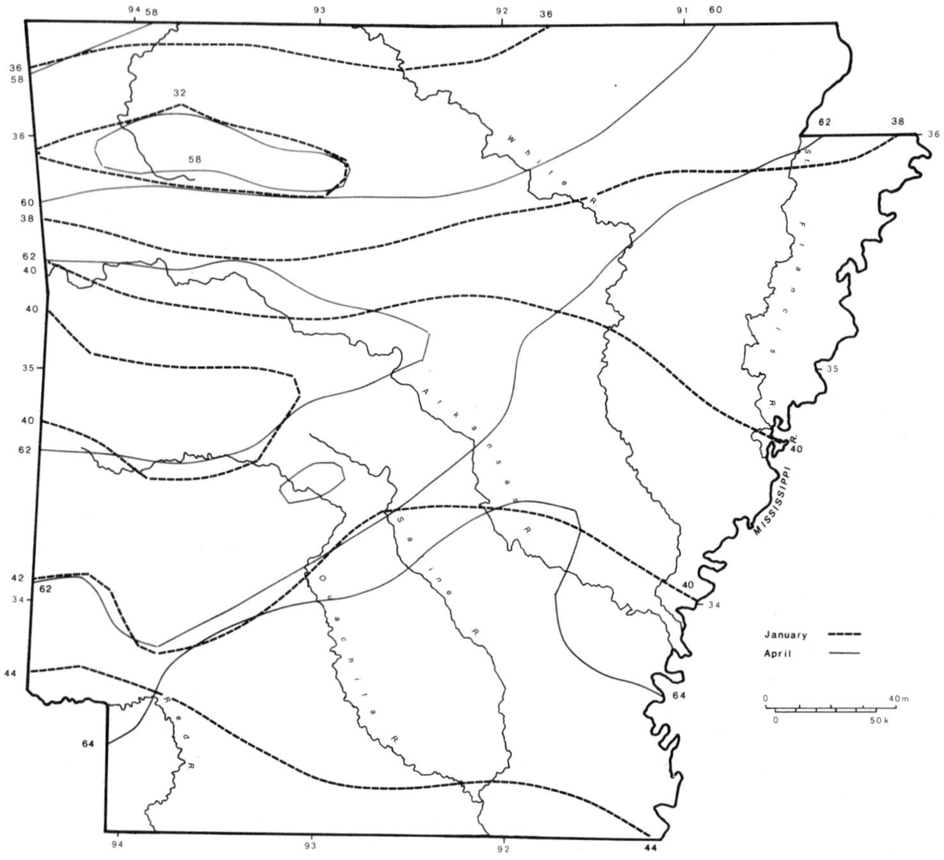

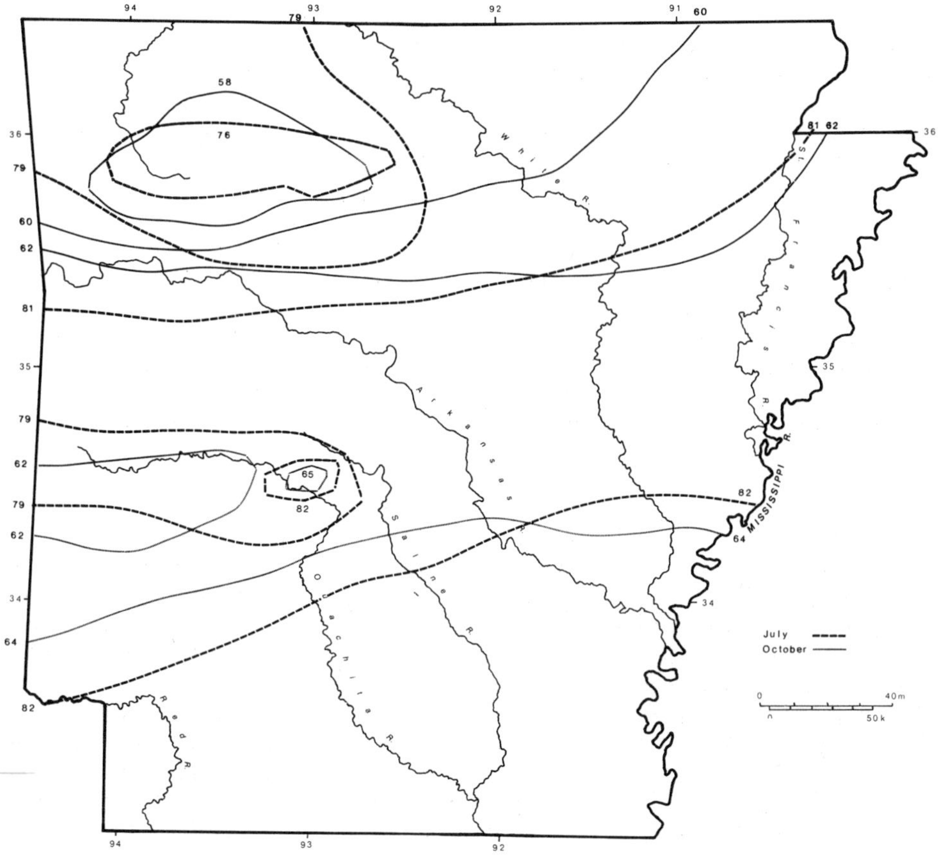

5. TEMPERATURE PATTERNS

5. TEMPERATURE PATTERNS

Arkansas's climate is classified as humid subtropical. Summers in the state are hot and humid; winters are short, cool, and mild, although brief periods of extreme cold do occur. Diurnal temperature differences are very small in summer and only moderate during the winter. The uplands experience slightly cooler temperatures than the lowlands.

Arkansas's location north of the Gulf of Mexico and inland some three hundred miles produces many humid subtropical features. The generally mild temperatures in winter and the hot weather in summer are direct results of the latitude and of the maritime tropical air masses that flow into the state from the Gulf of Mexico. These tropical air masses bring with them high humidity and high temperatures during the summer months and allow minimal cooling at night, making diurnal temperature ranges very low. Although temperature ranges are greater during the winter, because of occasional outbreaks of cool, dry air from the interior United States, winters are mild.

Summer average monthly temperatures are above seventy-five degrees Fahrenheit across the state. The dominance of high pressure, with considerable sunshine, spells hot weather; and the high humidity that accompanies the high pressure makes for considerable discomfort for people living in the area. Eight or more months during the year have average monthly temperatures over fifty degrees Fahrenheit. Only in December, January, and February does the monthly average drop below fifty degrees Fahrenheit. Northwest Arkansas experiences an average of thirty-four degrees Fahrenheit during those months, while the remainder of the state experiences monthly average temperatures between thirty-four and forty-two degrees Fahrenheit.

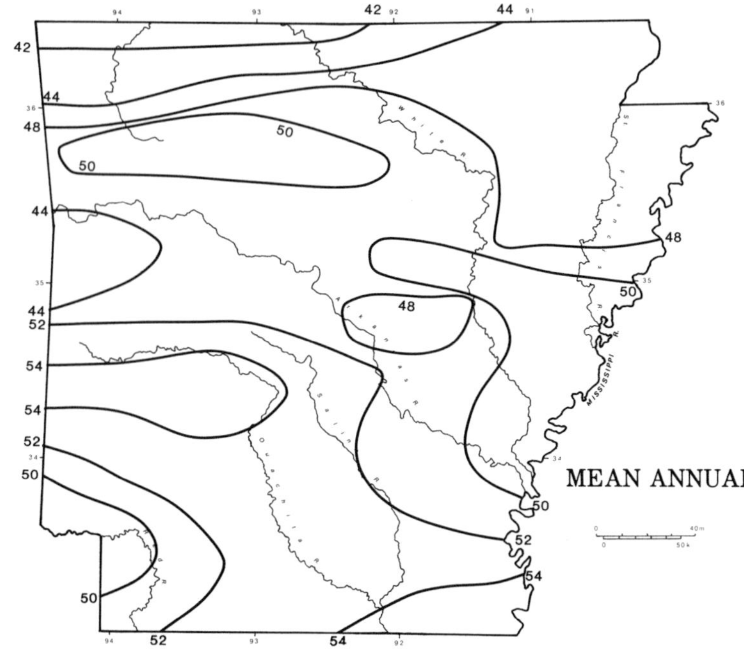

MEAN ANNUAL PRECIPITATION

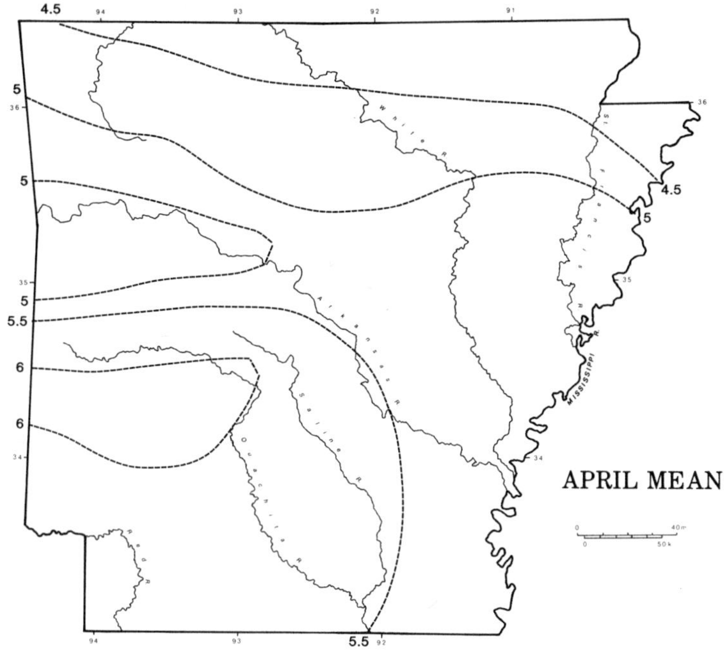

APRIL MEAN PRECIPITATION

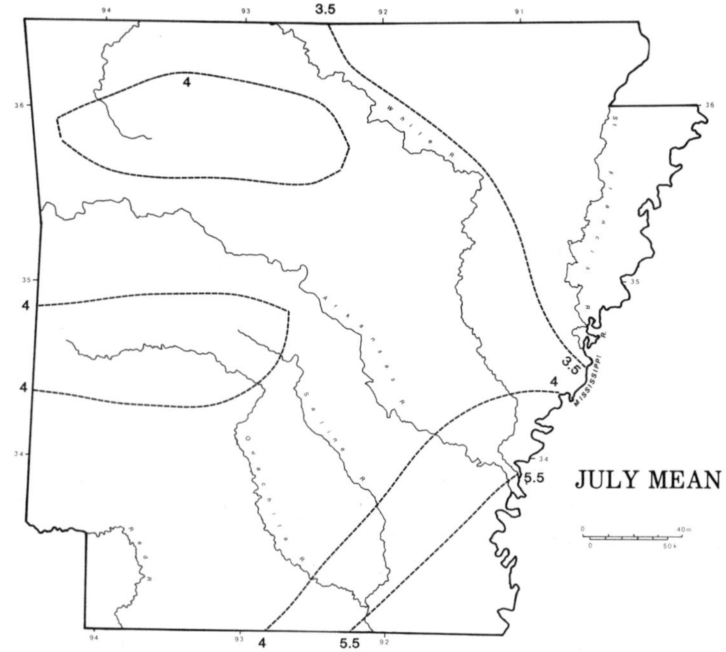

JULY MEAN PRECIPITATION

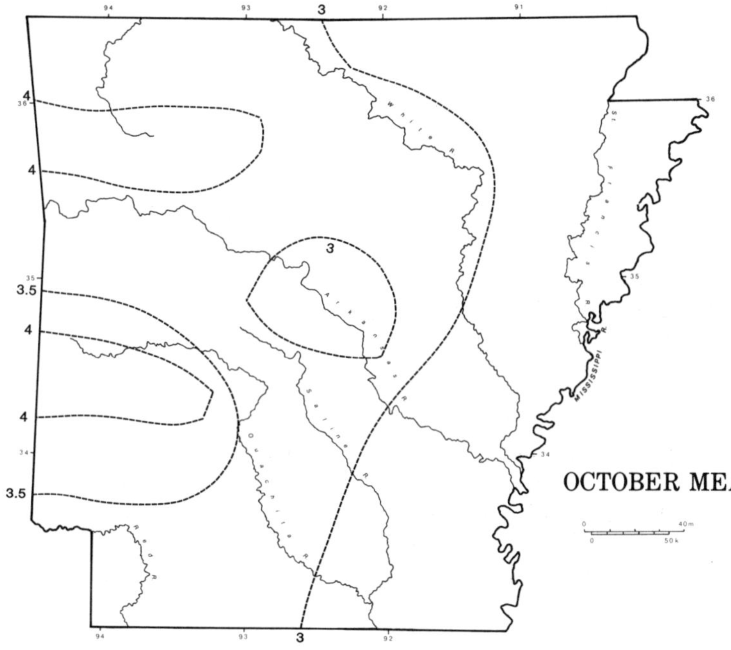

OCTOBER MEAN PRECIPITATION

6. PRECIPITATION PATTERNS

6. PRECIPITATION PATTERNS

Precipitation in Arkansas is most often in the form of rain. The moisture generally comes from the warm, moist maritime tropical air masses that originate in the Gulf of Mexico and move across the state during most of the year. Some seasonal variation occurs, but each month receives some precipitation.

Generally more rain falls in the late autumn, winter, and early spring, but precipitation is abundant and well distributed throughout the year, assuring well-sustained agricultural production. The average number of days of measurable precipitation is near one hundred. Only during the summers are there periods of infrequent rainfall, and then local showers occur in scattered locations.

During the winter frontal precipitation occurs as systems from the west and northwest clash with the warm, moist Gulf air. These clashes often produce heavy precipitation. In the summer convective processes dominate rainfall patterns. Rising warm air is heated from below by the hotter land surface. As this moist air rises, it cools and rainfall occurs in a local area. In highland areas orographic precipitation happens as the air meets landmasses. The cooling process produces precipitation.

Annual precipitation reflects Arkansas's geographic location and local orographic influences. Because they are nearer to the Gulf of Mexico moisture source, the southeastern counties receive more rain than the others. While precipitation averages are forty-four inches to forty-eight inches per year in the northwestern counties, the southeastern counties get more than fifty-four inches. Yet, some highland stations in the Ozarks and the Ouachitas receive more rain than the surrounding valleys because of the orographic effect of the mountains.

Winter and spring are the wettest times of the year. December and January produce the greatest amount of precipitation in the southern counties. Precipitation is highest in northern counties between March and May.

Snow occurs throughout the state, although principally in the northwest. The average snowfalls range from over a foot in the higher Ozark elevation to one to two inches in the southeastern counties.

Despite the generally abundant rainfall, short periods of dry weather are frequent over small areas of the state, and occasionally larger areas of the state experience severe droughts of longer duration. Notable periods of severe drought occurred in 1930, 1954, 1963, and 1980.

MEAN ANNUAL FROST-FREE DAYS

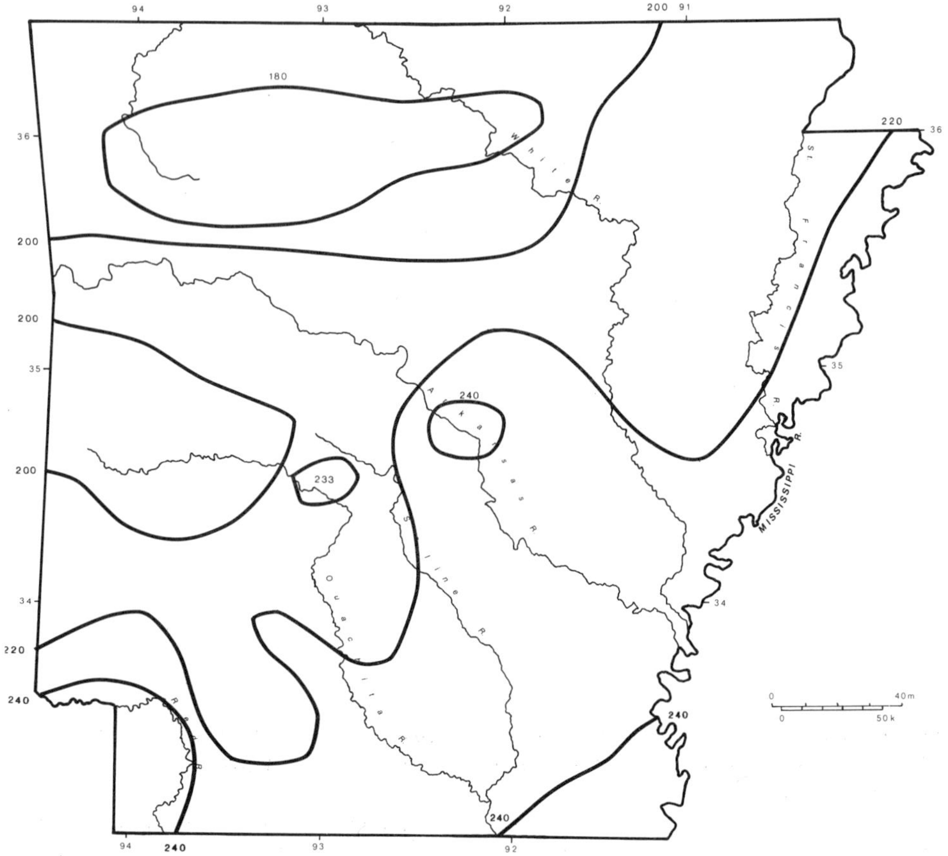

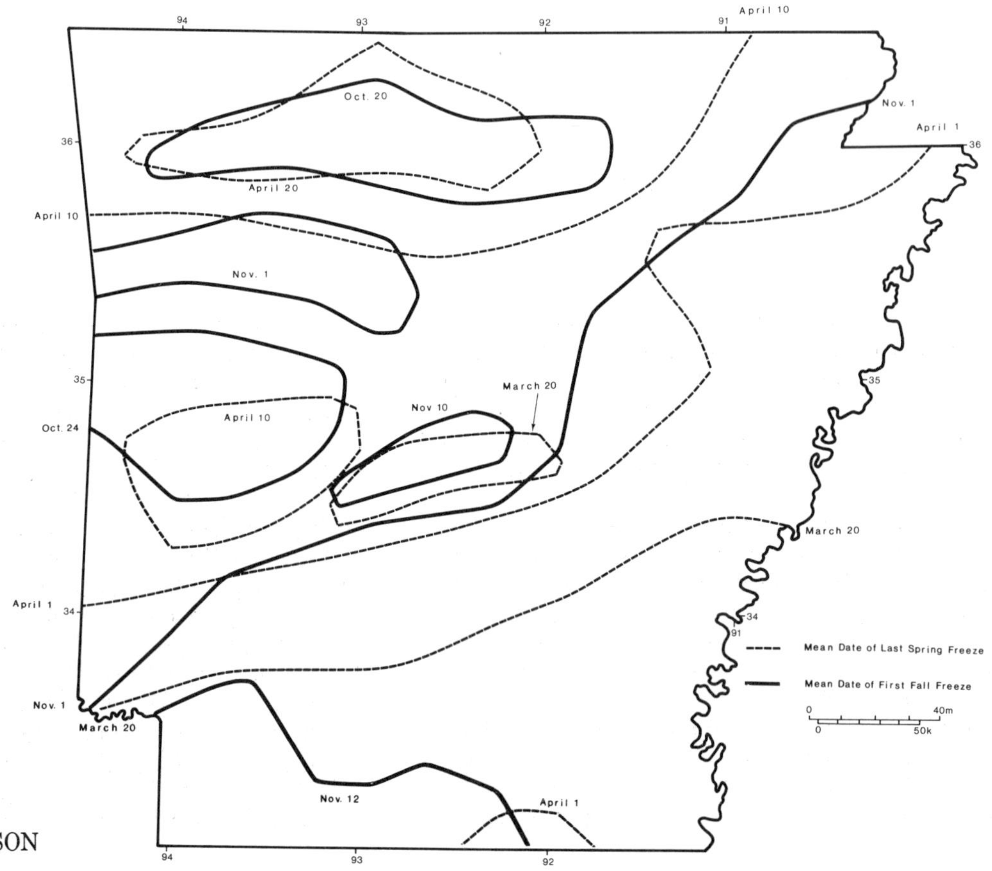

DATES OF THE FROST SEASON

7. GROWING SEASON AND FROST DATES

7. GROWING SEASON AND FROST DATES

The growing season is the period between the last killing frost in the spring and the first killing frost in the fall. Arkansas's growing season increases from the northwest to the southeast. There are, on the average, over 240 frost-free days in the southeastern quarter of the state. As one crosses into the uplands of the Ouachitas and the Ozarks, the season decreases to less than 200 days. Although the highlands are only a few hundred feet above the lowlands, the difference is enough to extend the frost season by several days. The Arkansas Valley, at a lower elevation than the highlands on the north and south, is a corridor with a longer growing season. The metropolitan center of Little Rock shows up as a prominent "heat island," where the built-up environment produces a frost-free season of over 240 days.

In addition to elevation differences, latitudes also play an important role in determing the frost-free season. The average dates of the last killing frost in the spring range from March 20 in the south to April 20 in the Boston Mountains. Given the elevation differences and the latitudinal variation, this substantial deviation is to be expected. The initial killing frost occurs about November 12 in the south and as early as October 20 in the north.

Historically, the growing season has played a major role in crop production. Optimum conditions for growing commercially important crops were found in the areas with over 220 frost-free days. For example, cotton was grown all over the state, but the longer growing season in eastern Arkansas made production easier there than in the highlands.

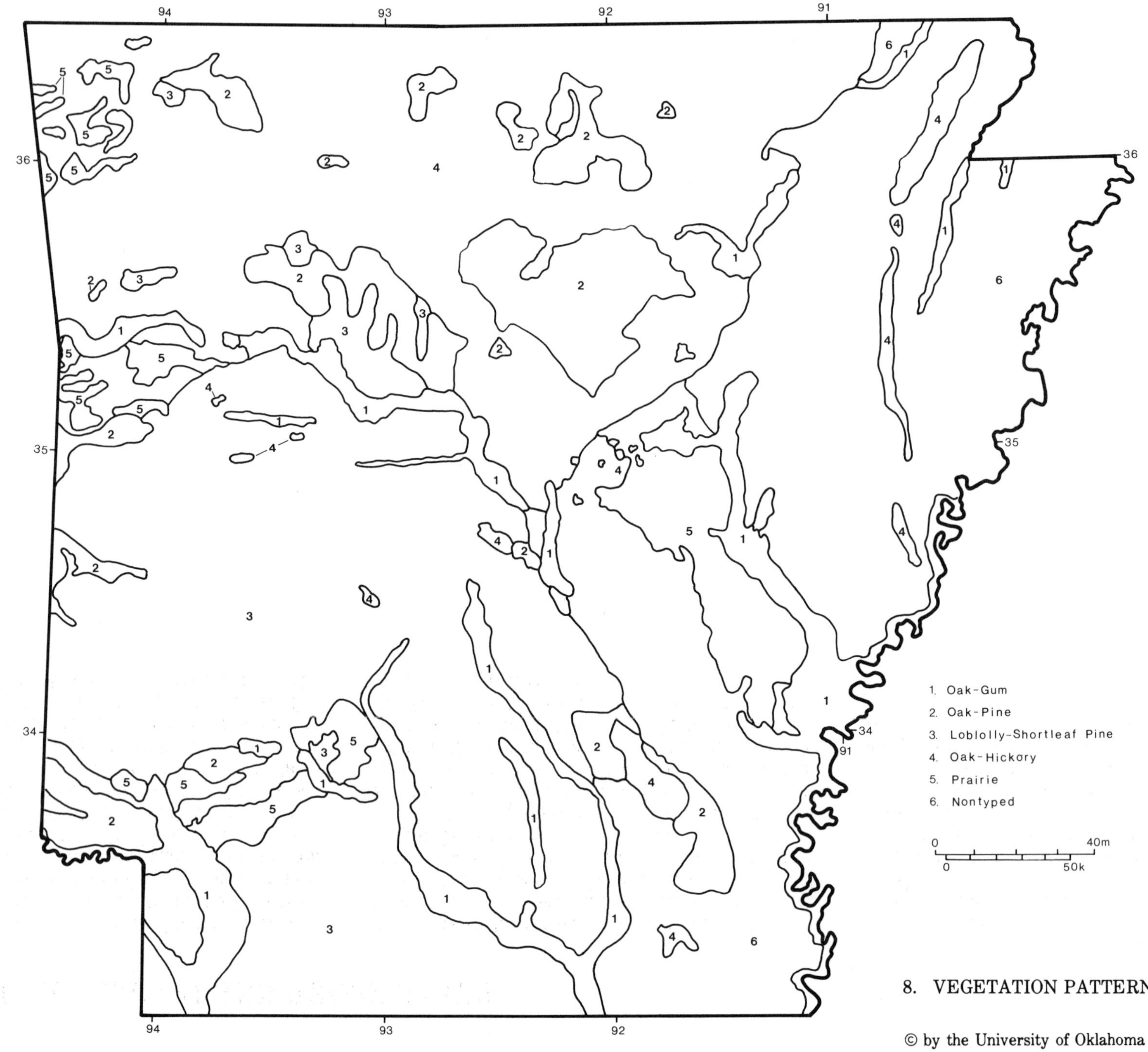

8. VEGETATION PATTERNS

8. VEGETATION PATTERNS

Five major categories of natural vegetation have existed in recorded times in Arkansas. While all have faced considerable pressure from human activity, they continue to survive. A sixth category, the nontyped system, is a modern phenomenon.

Gum, oak, tupelo, or cypress trees in the oak-gum forest (1) constitute at least 50 percent of the timber. Less than 25 percent is southern pine; cottonwood, willow, ash, elm, and maple are other associated trees. Generally this vegetation is found in the bottomlands of lowland Arkansas. These trees usually accompany standing water or seasonal water. Flood plains always contain these species. Along the Arkansas River and the White River in eastern Arkansas are extensive oak-gum forests. Originally such forests covered a much wider area in eastern Arkansas, but much of the stand there has been replaced by agriculture.

The oak-pine forest (2) has over 50 percent hardwood trees, mostly upland oak, gum, and hickory. Pines account for 25 to 50 percent of the timber. Growing environments include the southwestern corner of the state, a portion of the Alluvial Plain, south of the Grand Prairie, and scattered but large areas of the Ozarks. Some additional stands exist on the Blackland Prairie soils in southwestern Arkansas. Isolated stands also occur in the western Ouachita Mountains.

The most dominant forest group in Arkansas is the loblolly-shortleaf pine forest (3). More than 50 percent of the trees in these forests are varieties of the southern-pine species. Common associates include oak, hickory, and gum. The pines flourish in the sandy, infertile mountain and coastal soils of central and southern Arkansas. The loblolly dominates the area south of the Ouachita Mountains. The shortleaf species occurs more frequently between the Ouachita and the Arkansas rivers. The northern forest contains larger numbers of oak and hickory species.

The northern half of upland Arkansas is covered by an oak-hickory forest (4), where over half of the trees are oak and hickory species. Southern pine and red cedar account for less than 25 percent of the trees. Associated with the oak and hickory stands are gum, poplar, elm, and maple. The thin, moderately fertile soils of the Ozarks support this vegetation system adequately. Today the hardwood species are being cut but not replaced, and in their place southern pine is encroaching, giving rise to a change in forest species throughout the Ozarks.

Prairie (5) vegetation exists in its natural state only in a few isolated areas in Arkansas, although formerly grasslands covered much of the present Grand Prairie in the eastern part of the state. Tall prairie grass once flourished in Arkansas, Prairie, Monroe, and Lonoke counties. Although the prairie grasses have now been replaced by agriculture, the fertile soils of this area could still support them. Other regions of former prairie include northwest Benton, eastern Sebastian, western Logan, and northern Hempstead counties.

The last category, the nontyped (6) systems, includes primarily domesticated plants such as cotton, rice, and soybeans. The natural vegetation has long since been replaced by these plants in a considerable portion of eastern Arkansas.

PART TWO
PRE-EUROPEAN SETTLEMENT PATTERNS

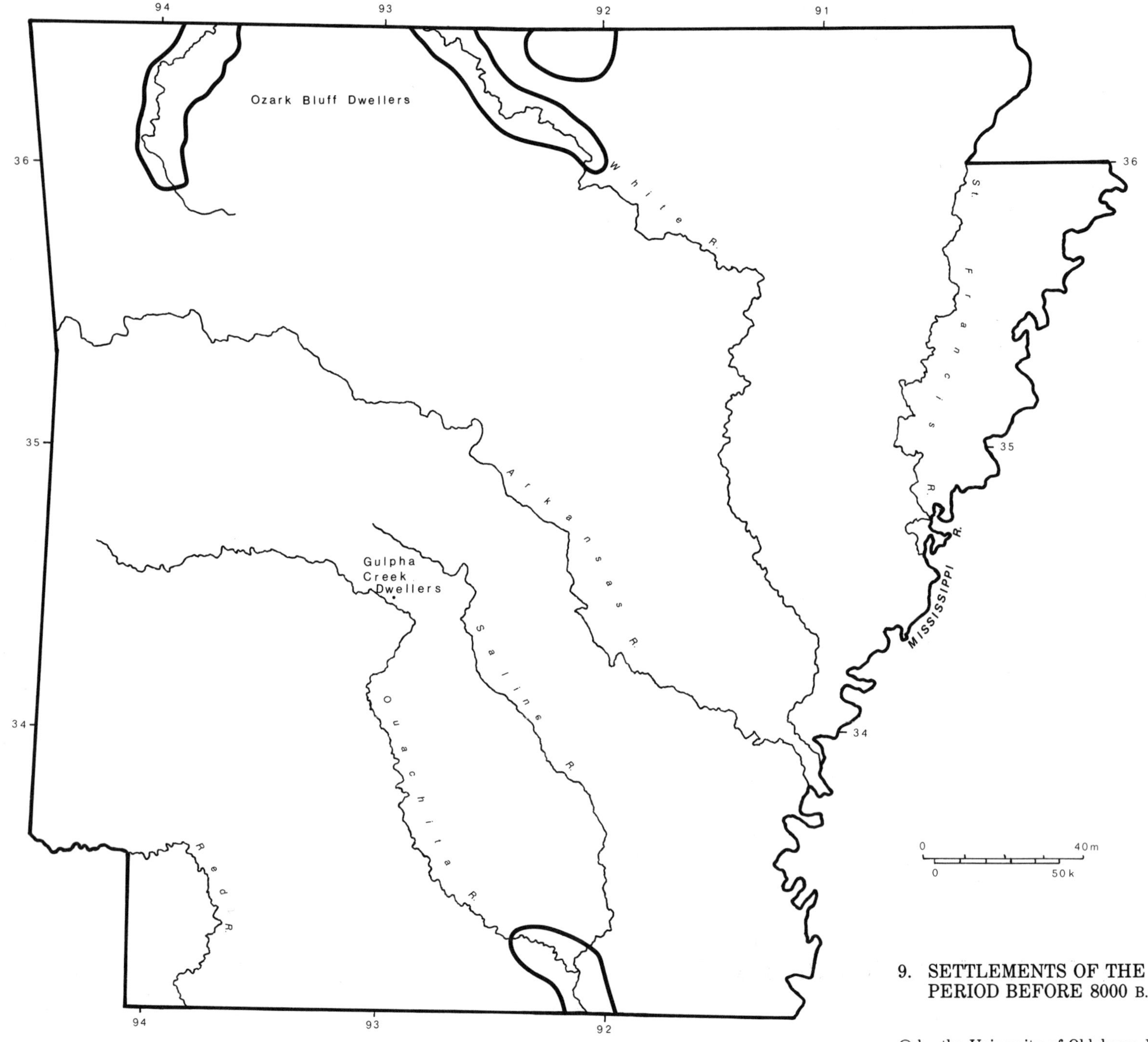

9. SETTLEMENTS OF THE PALEO-INDIAN PERIOD BEFORE 8000 B.C.

9. SETTLEMENTS OF THE PALEO-INDIAN PERIOD BEFORE 8000 B.C.

The earliest human inhabitants of what now is Arkansas were American Indians. A few sites indicate that they lived in the state as early as the Paleo-Indian period before 8000 B.C., sharing what appears to have been a common culture, although the evidence precludes precise differentiation among groups. Little remains of their culture, in fact, other than projectile points and stone tools.

Paleo-Indian sites, identified solely by the presence of distinctive spear and dart points, are found throughout much of Arkansas. The projectile points have been found in the bluff shelters of northwest Arkansas and along most of the river valleys of the state. Two principal locations are shown on this map for sites on the Ozark bluffs and Gulpha Creek. Fuller identification of the culture that produced the points has been made difficult because none of the sites from which these artifacts have come were well excavated, and indeed most of the points have been found on the eroded surface of the ground. Their presence indicates that the Indians of this early time period followed game through most of Arkansas.

Paleo-Indians in Arkansas produced their distinctive projectile points by percussion and pressure flaking, then attached them to cane or wood shafts for use as a thrusting weapon. The distinguishing feature of Arkansas and other Paleo-Indian projectile points is a central flute that was removed from the base of the point after the other flaking was finished. Other tools possessed by this culture included stone scrapers and knives and, undoubtedly, tools made of wood or other raw materials which have not survived.

Paleo-Indian tools have been found outside of Arkansas alongside the bones of various animals that were in the area towards the end of the Ice Age, including bison, mastodon, ground sloth, musk-ox, and tiger. Archaeologists assume that the same relationship existed in Arkansas, suggesting that Paleo-Indian groups were primarily hunters and gatherers. That these people were able to kill such large animals implies that they had achieved some level of cooperative society, although nothing is known about their communal life. They apparently had no permanent settlements, but roamed the country following the game upon which they lived.

9. SETTLEMENTS OF THE PALEO-INDIAN PERIOD BEFORE 8000 B.C.

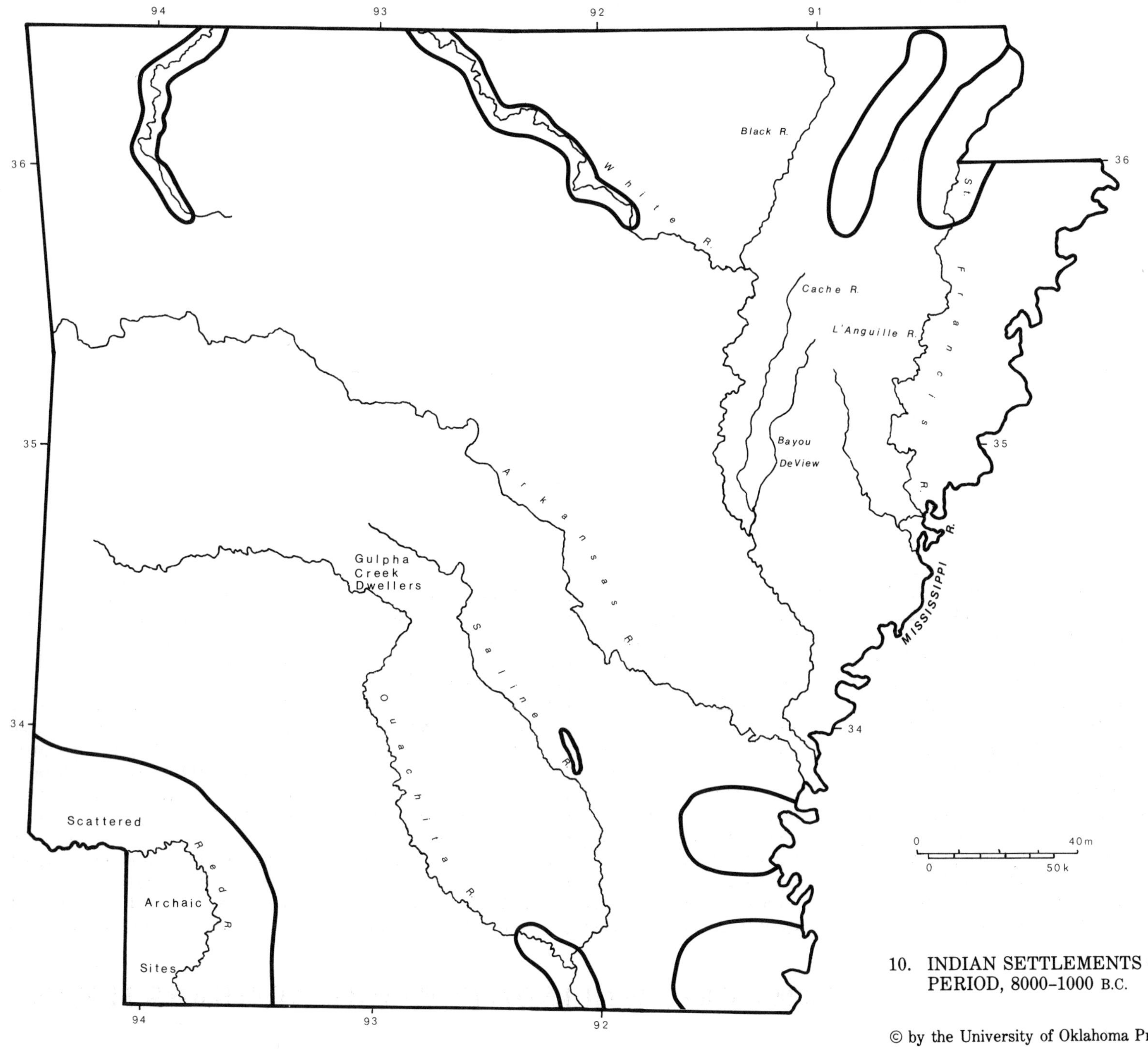

10. INDIAN SETTLEMENTS OF THE ARCHAIC PERIOD, 8000–1000 B.C.

10. INDIAN SETTLEMENTS OF THE ARCHAIC PERIOD, 8000–1000 B.C.

During the Archaic period from 8000 to 1000 B.C. the human population in Arkansas experienced growth and technological progress. Many Archaic sites exist in all parts of the state. Gulpha Creek, near Hot Springs, continued to be a place occupied by Indians, probably because of the existence of large quantities of flint, which was worked there into tools. The northwestern Archaic sites have been the best preserved, because of their protection from the weather, and consequently they provide the best evidence of Archaic culture.

Archaic Indian cultures are distinguished by changes in the shape and method of producing their projectile points and tools. In addition, they expanded the components of their basic tool kit. The spear and dart remained the major projectile points of these Indians, but they were no longer made with central flutes, and notches were produced at the base of points to make possible a more stable attachment of the point to the shaft. During the Archaic period flaking gave way to pecking (that is, using a hammerstone to beat off pieces of a softer stone) and grinding as the principal means of a shaping Indian tools. Besides spear and dart points, other tools associated with Archaic sites are adzes, axes, celts, hammerstones, mortars, and net sinkers.

Innovations in technology other than tools are also associated with Archaic Indians. At some unknown time they learned weaving, a technology that they used to produce baskets, nets, cloth, and cord out of cane and grass. Woven artifacts are found only in the bluff shelters of northwest Arkansas, but a similar culture may have been pervasive, although not preserved, among contemporary groups throughout Arkansas.

These Indians' basic diet also changed, perhaps as a result of the disappearance of the large game animals hunted in the Paleo-Indian period. Archaic trash middens show that they consumed deer, modern bison, small game, fish, mussels, nuts, berries, and other plants.

Knowledge of the society of Archaic Indians remains largely speculative. They were still indistinguishable bands of hunters and gatherers, although the existence of different styles for projectile points indicates the emergence of cultural variation. They remained wanderers with no permanent village sites, although the first evidence of what may be stylized, formal burial practices is from this period. The dead were buried in shallow grass-lined pits with their knees and hands bent to place them under the chin. This evidence suggests the emergence of more complex religious or political structures, although the nature of these structures remains hypothetical.

10. INDIAN SETTLEMENTS OF THE ARCHAIC PERIOD, 8000–1000 B.C.

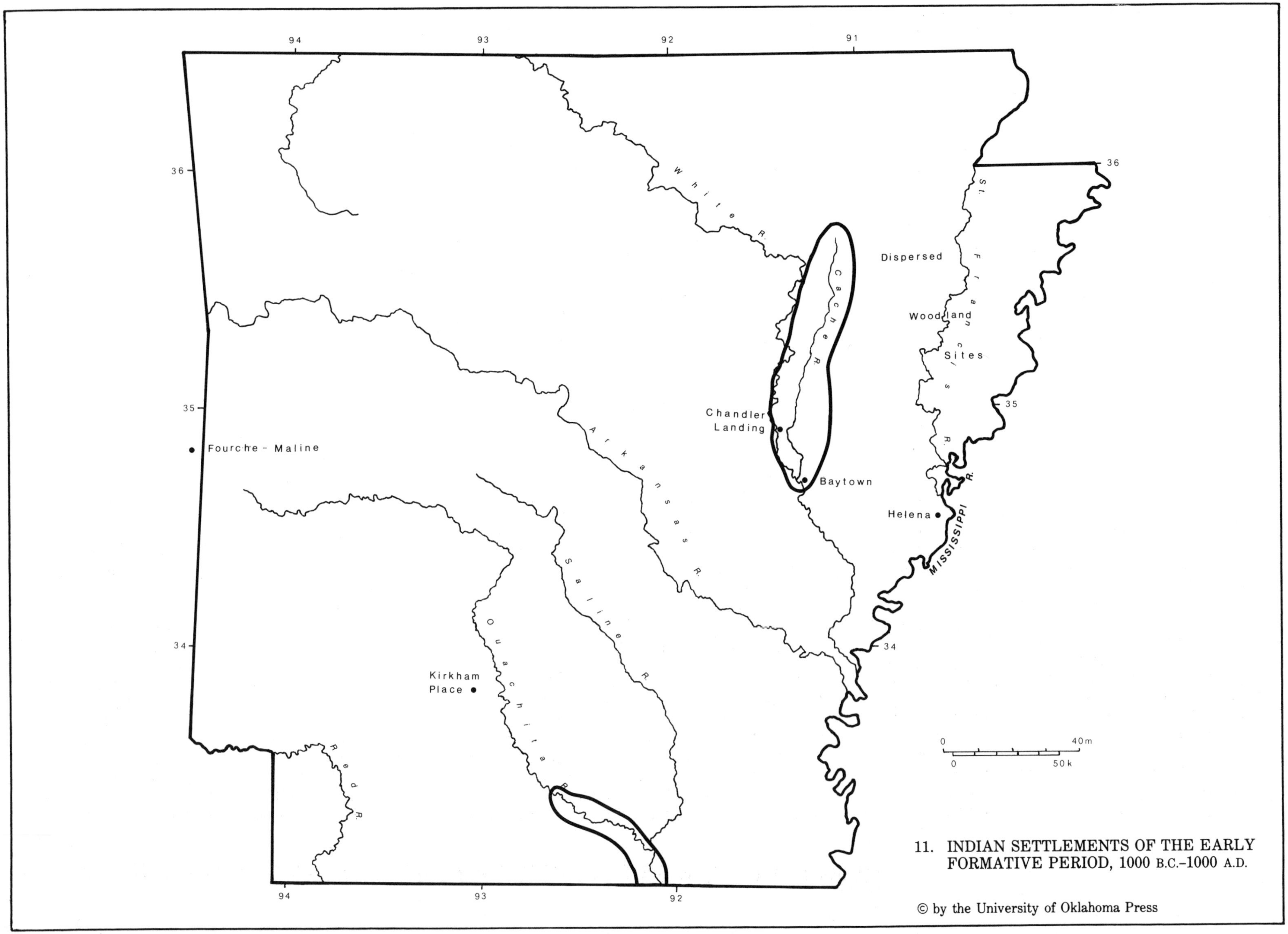

11. INDIAN SETTLEMENTS OF THE EARLY FORMATIVE PERIOD, 1000 B.C.–1000 A.D.

11. INDIAN SETTLEMENTS OF THE EARLY FORMATIVE PERIOD, 1000 B.C.–1000 A.D.

The Formative period in Arkansas began about 1000 B.C. and ran into the historic era, which began in the middle of the sixteenth century A.D. Archaeologists separate the Formative period into early and late phases, the former ending about 1000 A.D. Early formative culture in Arkansas is associated with the Woodland Indian cultural development and is sometimes called the Burial Mound period. It was a revolutionary phase in the expansion of Indian culture, marked by the introduction of agriculture and pottery making, the development of permanent village life, and the construction of burial mounds.

The source of the agricultural expertise among these Indians in Arkansas is uncertain. Probably the technology spread into the area from Mexico. By the early formative period, however, Indian groups in the Arkansas area grew the basic crops found among similar cultures eleswhere, including Indian corn and squash.

The introduction of agriculture did not end hunting, and the Indians continued to supplement their new diet with game. In fact, hunting was made easier by a major technological innovation, the introduction of the bow and arrow. Projectile points capable of being used on arrows were now made for the first time.

Pottery may have entered this area from Mexico as well. Most pottery from Arkansas sites of the early Formative period was undecorated. Where decoration existed, it was either in the form of indentation of the wet clay with cord-wrapped paddles or the incision of lines made with shells or other tools.

Agriculture made possible the food supply necessary for a more sedentary life among the Indians. As a result, the first permanent villages emerged during the early formative period. Sites were widely scattered, although settlements were particularly concentrated on the delta in eastern Arkansas, with numerous sites along the upper Mississippi and others along the Cache River in northeast Arkansas. In addition, there were sites along the Red River in the southwest. Major sites include a Helena focus, in the vicinity of that town on the Mississippi; Baytown, in Monroe County; Chandler Landing, in Prairie County; Kirkham Place, on the Little Missouri in Clark County; and Fourche-Maline, on a creek by that name near its junction with the Poteau River, just west of the present-day Oklahoma-Arkansas border. During the Formative period the villages in the Ozark and Ouachita mountains were much smaller, and the population did not increase as much as elsewhere, although there is good evidence that corn, sunflowers, squash, and other plants were being cultivated on all the creek and river bottoms.

The character of early Formative society remains conjectural to a great degree. The most obvious manifestation of an expanded social organization was in the society's treatment of the dead, who were no longer buried in pits, but placed on the top of the ground or in shallow graves with earth mounded over them. Typically, all village burials became confined to such mounds, with many graves being placed in the same area. While this evidence indicates the presence of a more complex religious life and collateral social organizations, the specific nature of this religion and society is not clear.

11. INDIAN SETTLEMENTS OF THE EARLY FORMATIVE PERIOD, 1000 B.C.–1000 A.D.

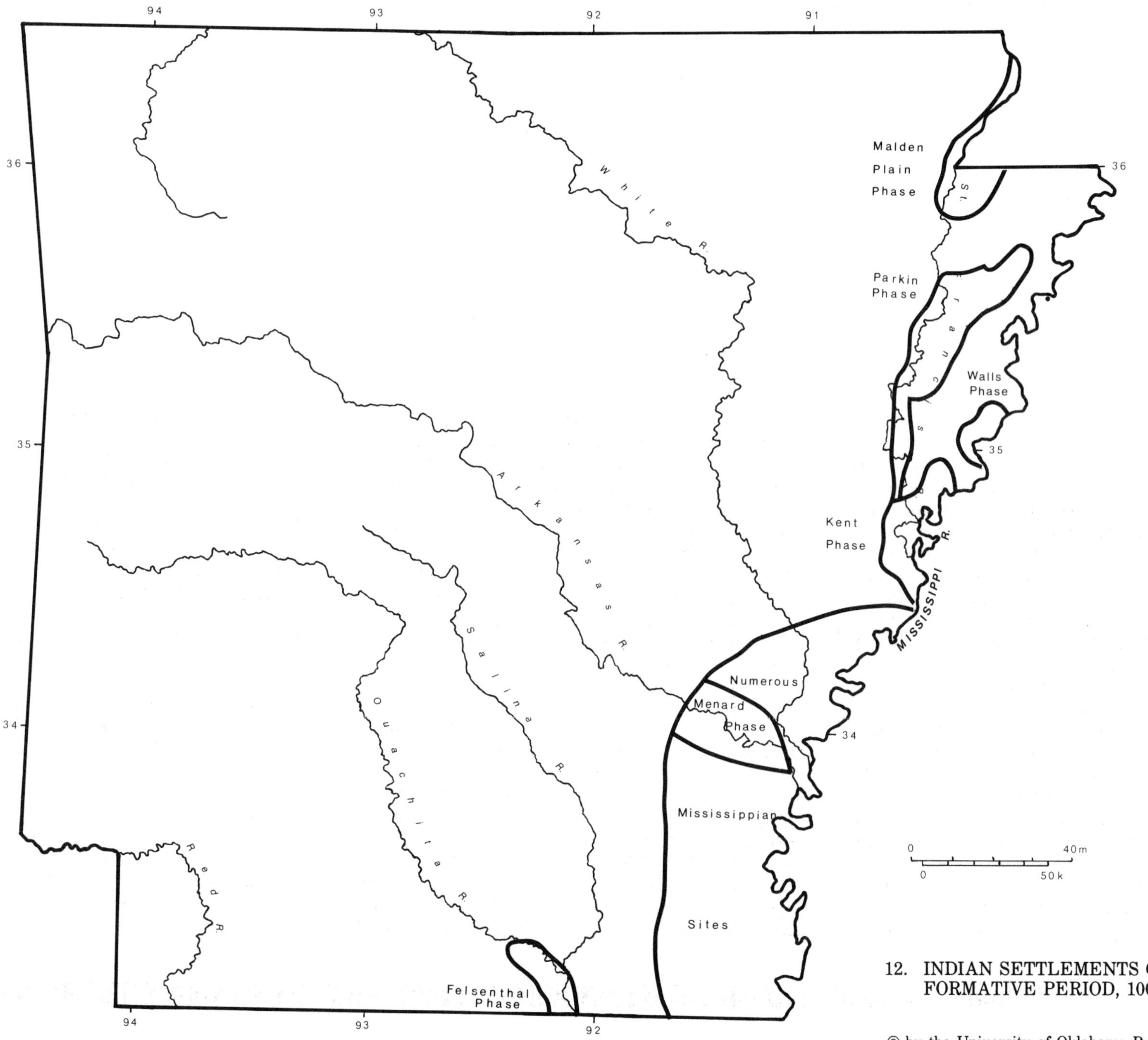

12. INDIAN SETTLEMENTS OF THE LATE FORMATIVE PERIOD, 1000 A.D.–1500 A.D.

12. INDIAN SETTLEMENTS OF THE LATE FORMATIVE PERIOD, 1000 A.D.–1500 A.D.

In the late Formative, or Temple Mound, period, 1000 to 1500 A.D., various sites emerged that are associated with what is called the Mississippian stage of cultural development. There was considerable activity in the southwest, southeast, and northeast of Arkansas, where major communities were developing in the nonmountainous portions of the state. These groups continued the village tradition established earlier.

This period was marked by cultural differentiation among the emerging communities. The areas south of the Arkansas River received their greatest cultural influence from Louisiana and southwestern Mississippi, yet these people were further divided by cultural contacts. Those between the Mississippi and Ouachita rivers were affiliated with the cultures of the Mississippi River valley, and those west of the Ouachita with the Caddo culture of the southwest. In the north, the peoples on the north and east had greater impact.

The clearest evidence of regional cultures is seen in the pottery of late Formative Indians. The Indians north of the Arkansas used crushed mussel shell to temper their pottery. Indians south of the river used dry clay or crushed bone. Decoration also served to differentiate the two regions. In the north pottery was incised (that is, decorated while still wet), while in the south it was engraved (cut after the pottery was baked or hardened). Red and white paint was used on pottery north of the river, but not typically in the south.

Late Formative societies included several archaeologically distinct phases. The Parkin phase is represented by sites along the Saint Francis River, including Parkin. Settlement clusters along the Mississippi River between present-day West Memphis and Helena comprise the Walls, Malden Plains, and Kent phases. The Menard phase, a subculture strongly influenced by Caddoan culture in the south, was concentrated in the lower Arkansas River valley. Various distinctive foci and the Felsenthal phase along the lower Ouachita River constituted the Formative population of southern Arkansas. Sites along the Red and Little rivers are part of the late Formative culture in the southwest.

Among all of these groups, agriculture became more important as populations grew. Population pressure required a greater emphasis on sedentary agriculture. Agricultural products were still supplemented, however, with wild game.

The late Formative period saw an increase in the number of tools used by these cultures. In this era the usual stone tools were supplemented by many implements made from bone, including arrowpoints, awls, and combs.

The presence at living sites of items clearly not available in Arkansas suggests the growth of trade among regional Indian groups. In particular, sea shells, used as ornaments or drinking vessels, indicate the distances such trade must have covered.

The late Formative period was characterized by the emergence of permanent village complexes. In eastern Arkansas these were large, covering up to twenty acres. Permanent houses and a ceremonial pyramid were commonly a part of the villages. Some of the villages were fortified. The Parkin site on the Saint Francis River is representative, with a moat running around the entire edge of the village and a wooden palisade built inside the moat. This type of structure indicates considerable interaction among Indian groups during the period, some of possibly violent.

12. INDIAN SETTLEMENTS OF THE LATE FORMATIVE PERIOD, 1000 A.D.–1500 A.D.

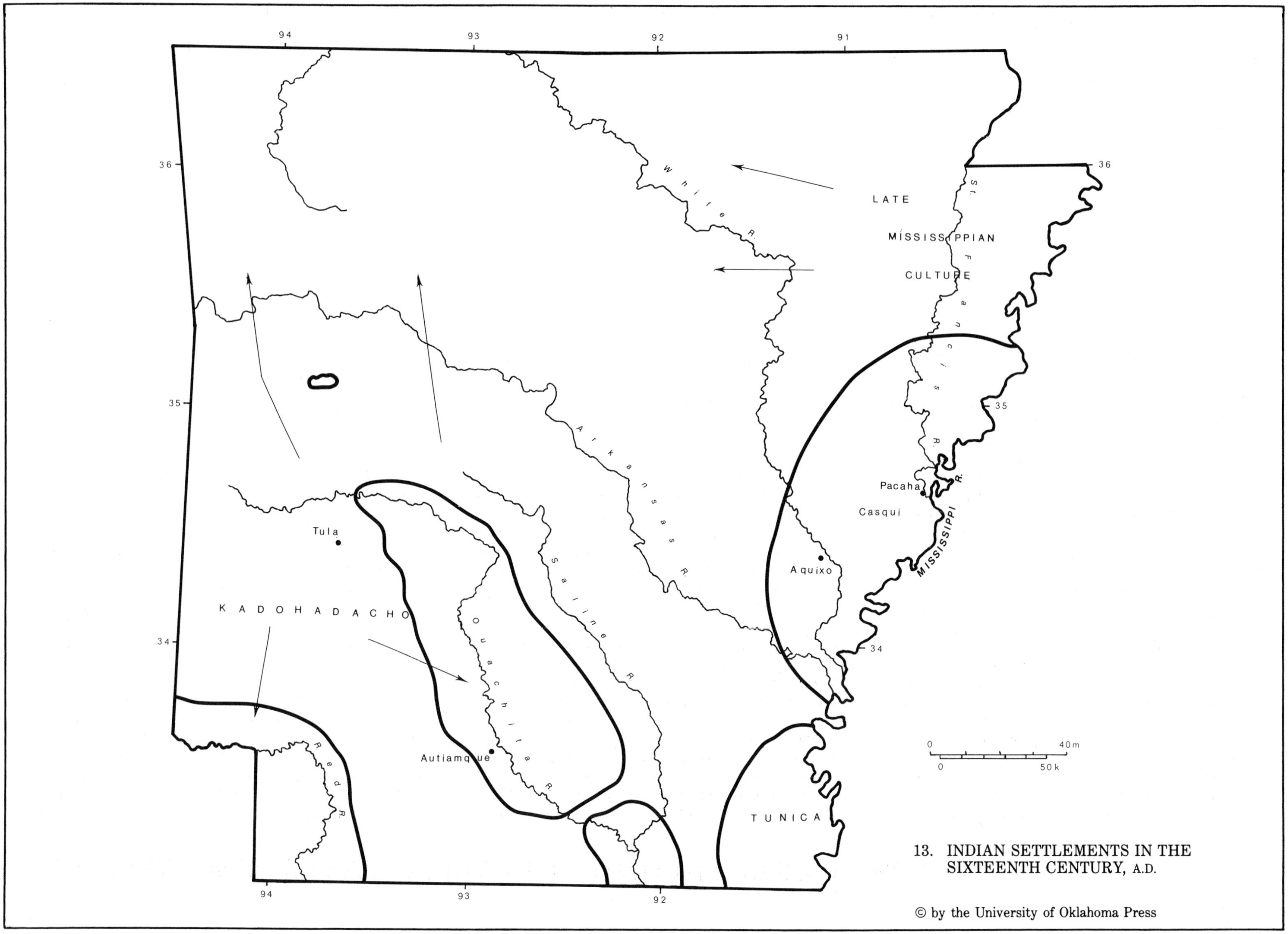

13. INDIAN SETTLEMENTS IN THE SIXTEENTH CENTURY, A.D.

13. INDIAN SETTLEMENTS IN THE SIXTEENTH CENTURY, A.D.

The first historical reports of the Indians of Arkansas appeared after Spanish contacts in the sixteenth century. In 1541 the expedition of Hernando de Soto moved through the region. De Soto reported a large Indian population inhabiting the areas that he explored along the Mississippi, Arkansas, and Ouachita rivers. While the reports of his expedition do give a basic overview of the Indians of sixteenth-century Arkansas, the observers failed to provide enough details to link precisely the people they found with later Indian groups.

De Soto encountered Indians all along the Mississippi after landing near modern-day Helena, Arkansas. Marching northward (see map 21), he went through towns that he called Aquixo and Casqui on his way to a province called Pacaha, the principal town of which was a fortified village also known as Pacaha. Pacaha was probably located in the vicinity of the mouth of the Saint Francis River. Archaeologists once thought that this village contained the forebears of the Quapaw, who occupied this area later, but the historian W. David Baird suggests that the Quapaw had not yet reached Arkansas at this time. The linkage between Pacaha and historic Indians remains unclear.

Linguistic and cultural similarities convinced the Spanish that all of the Indian groups were related that occupied a territory running up the Arkansas River from its mouth as far as present-day Little Rock, and from the Arkansas across to the L'Anguille River. They were not politically unified, however, and de Soto attempted to use their hostility towards one another to his advantage. The Spanish explorer estimated that between six and seven thousand individuals lived scattered about that area. From the reports of the expedition it is clear that these people were in the late Formative stages of development, most of them living in fortified villages with timber stockades around them. Near these villages were extensive fields in which the Indians cultivated corn and beans. While they were agricultural, they also fished local streams, using nets, and hunted both small and large game.

When de Soto's expedition crossed the Ouachita Mountains, it encountered Indian groups that were different from those of the Arkansas River region. He met fierce resistance, near Caddo Gap, in the province he called Tula. The group that he fought at Tula may have been the Cahinnio tribe of the Kadohadacho Confederation which was in this area at about the time of the de Soto expedition. De Soto encountered other Indians who were equally hostile as he moved southward from Tula. He wintered with one group at a town he called Autiamque. Probably all of the Indian groups he discovered along the Ouachita and Little Missouri rivers were related to the Kadohadacho in some way. These Indians possessed permanent villages and extensive agriculture, cultivating in their fields squash, beans, pumpkins, and sunflowers. Their use of snares to capture rabbits attracted Spanish curiosity.

De Soto also contacted several other Indian groups in southeast Arkansas. They have not been identified definitely with any historically recorded tribe, but may have included the Tunica, the related Koroa, and prehaps a small northern-Louisiana group called the Taensa.

13. INDIAN SETTLEMENTS IN THE SIXTEENTH CENTURY, A.D.

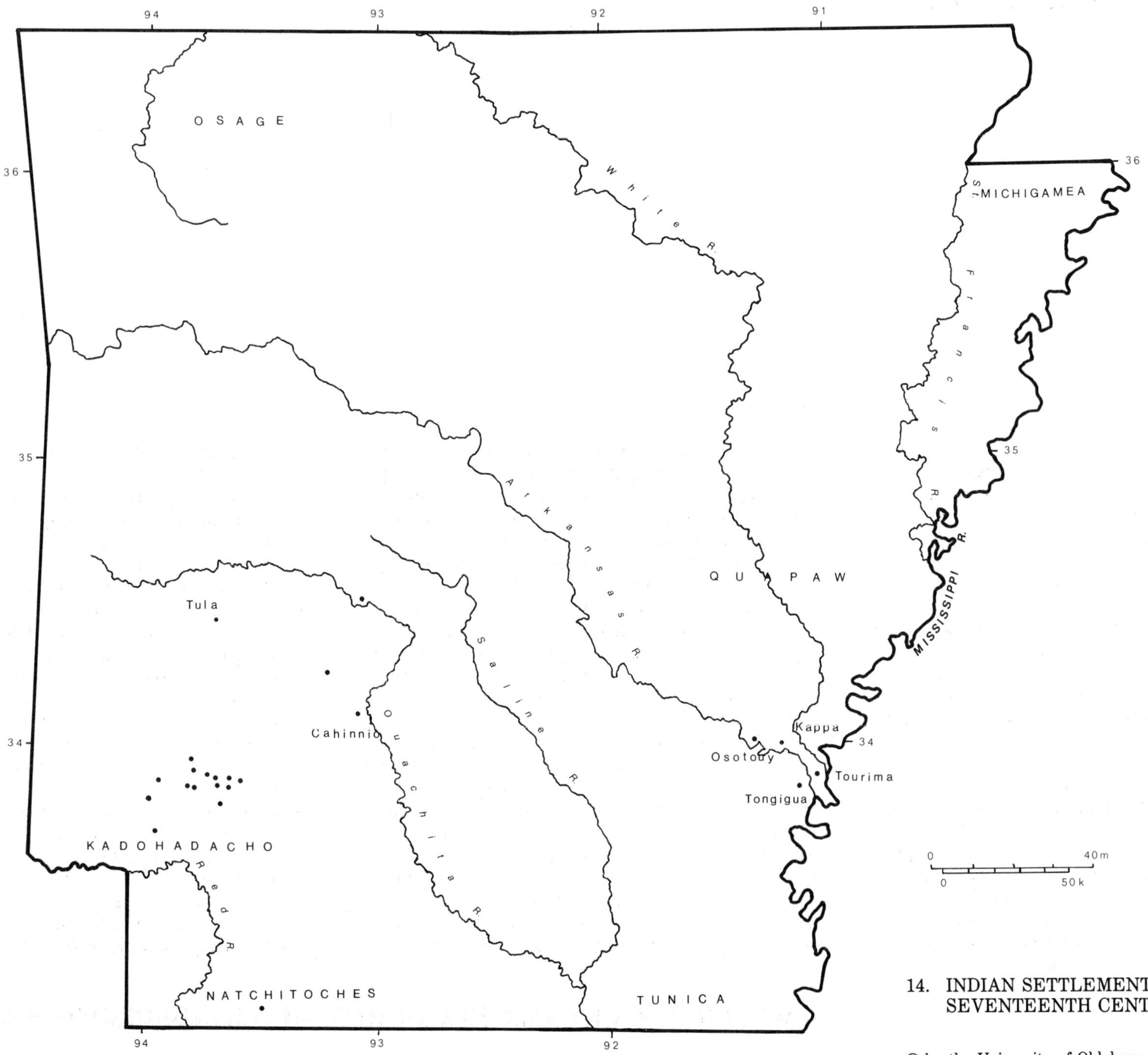

14. INDIAN SETTLEMENTS IN THE SEVENTEENTH CENTURY, A.D.

14. INDIAN SETTLEMENTS IN THE SEVENTEETH CENTURY, A.D.

Between the time of de Soto's contacts in 1541 and those of the French in 1673 the Indians in Arkansas changed considerably. Population appears to have declined. In place of the widespread groups of Indians found by the Spanish, the French encountered only a few. Louis Joliet and Père Marquette, for example, found only two Indian villages along the Mississippi, where de Soto had found many. There are two possible reasons for the different observations. First, there may have been greater concentrations of population as the agricultural revolution spread, placing larger groups of Indians in population centers and making them less dispersed. Second, there may have been an actual population decline because of diseases introduced into the Indian population by the Spanish over a hundred years before. For whatever reason, the French found real changes in the Indian population.

The first major group encountered by the French were the "Arkansea," observed by Marquette and Joliet in 1673. On his expedition in 1682, La Salle found the same people in four villages and called them the "Kapaha," or Kappa. These Indians were the Quapaw of later historical times. The historian W. David Baird believes that their name meant "downstream people" and that they may have moved into the area around the mouths of the Arkansas and White rivers sometime after the de Soto expedition, although archaeologists believe that the evidence places these Indians in Arkansas before de Soto. Linguistically these people were related to the Omaha, the Kansa, the Ponca, and the Osage on the Missouri River. The Quapaw possessed a basic agricultural technology, but also were hunters and gatherers. They resided in permanent villages, living in cabins made of bark. Although their origins are not clear, their major cultural influence appears to have been out of the northeast. Yet they also engaged in pottery trade with the Caddo on the south. There were at least four major Quapaw villages: Tourima, west of the junction of the White and Mississippi rivers; Tongigua, above the mouth of the Arkansas; Kappa on the Mississippi north of the White River; and Osotouy, farther up the Arkansas River.

Marquette also located the Osage in modern-day northwest Arkansas during his 1673 expedition. This tribe had moved into northern Arkansas at approximately the same time that the Quapaw had moved to the mouth of the Arkansas River. The Osage were an agricultural group with villages in central and southwestern Missouri, but they engaged in hunting that took them into Arkansas, following the migrations of buffalo and other game animals. There is no archaeological evidence of Osage communities in Arkansas, but they definitely ranged into the area and were frequently at war with the Quapaw and the Caddo in the southwest.

Henri Joutel and the survivors of the La Salle expedition visited the Caddo Indians in southwestern Arkansas near modern-day Fulton. They found the group known as the Kadohadacho, one of three Caddo confederations. These Caddos, descendants of previous Indian groups in the area, were hunters and farmers. They cultivated traditional Indian crops such as corn, squash, beans, pumpkins, and sunflowers. Permanent Caddo settlements were located along many of the streams of southwestern Arkansas. The French estimated that there may have been as many as eight thousand Caddo in the three confederations.

Among the lesser Indian groups found by the French, the Tunica had moved north along the Mississippi and had villages in the southeastern part of the state. On the south the Natchitoches Caddo Confederation had extended north into Arkansas, and a subdivision, the Ouachita, were settled along the Ouachita River near present-day Columbia County. The Michigamea, a division of the Illinois, were located along the Mississippi, either in northeastern Arkansas, near Big Lake in Mississippi County, or, according to Samuel D. Dickinson, near the mouth of the Saint Francis River in Phillips County.

14. INDIAN SETTLEMENTS IN THE SEVENTEENTH CENTURY, A.D.

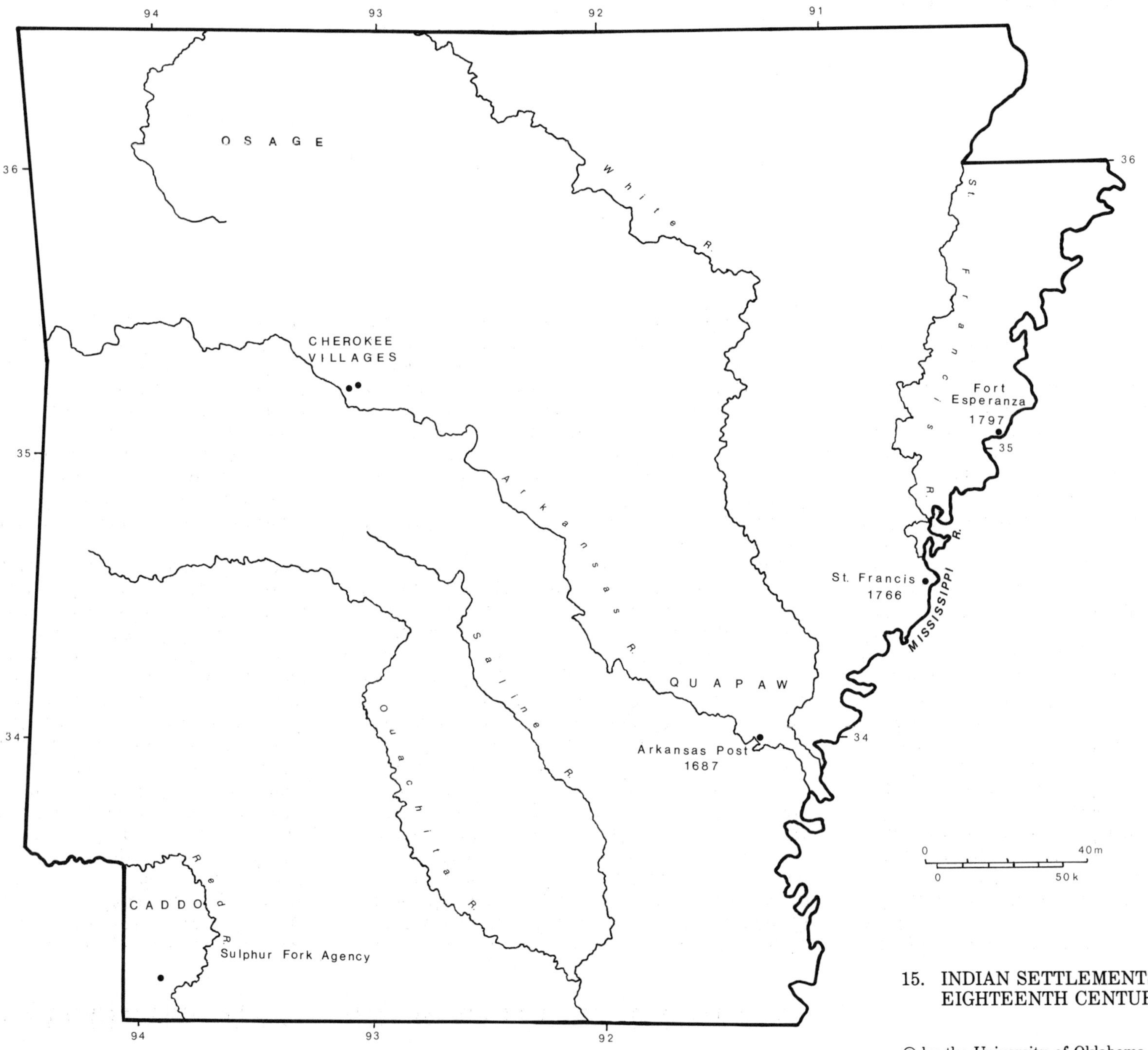

15. INDIAN SETTLEMENTS IN THE EIGHTEENTH CENTURY, A.D.

15. INDIAN SETTLEMENTS IN THE EIGHTEENTH CENTURY, A.D.

In 1800 the Indian groups identified by the French in the last quarter of the seventeeth century still survived, in Arkansas, but they faced increasing destabilization. The French contacts had not been extensive, but during the eighteenth century French, British, Spanish, and then American pressure had steadily increased. The frontier of the United States was pressing westward, and by 1800 the white population had begun to move into the Arkansas area. With this movement, waves of Indians were pushed out of the eastern United States into Arkansas, and afterwards the expanding white population pressed directly on the Indians in the Arkansas area.

The Osage were still active in the northwest. They raided the Caddo along the Red River in modern-day Louisiana, Arkansas, Oklahoma, and Texas. They also raided the Quapaw on the lower Arkansas River. According to one contemporary observer, the Osage were always at war with the world.

Yet the Osage control of northwest Arkansas was being challenged at the end of the eighteenth century by the Cherokee, who were moving westward out of the United States. The first Cherokees had moved into northeast Arkansas after 1790, and by 1817 there was a Cherokee settlement along the Arkansas River near present-day Russellville. Continued migration into Arkansas increased the Cherokee populaton to the point that they and the Osage were usually at war over territorial rights.

The Quapaw had moved farther up the Arkansas River by 1800, and visitors reported the tribe to be declining in population. In 1750 their population was estimated at about fifteen hundred. They were almost totally confined to four villages on the north and south banks of the lower Arkansas in the vicinity of Arkansas Post. This striking reduction in population and territory has not been explained, although disease and wars, with the Chickasaw across the Mississippi and the Osage on the north, are possible factors. In 1748 or 1749 the Quapaw had allied themselves with the French at Arkansas Post against the Chickasaw, who then lived in present-day northern Mississippi. This war was a major defeat for the Quapaw. After 1800 the Quapaw were the first tribe within the Arkansas area to confront pressure from growing numbers of white settlers, because they occupied one of the major immigration routes, the Arkansas River valley.

On the south the Arkansas Caddo had consolidated on either side of Red River at the Great Bend, where that stream turns south before entering modern-day Louisiana. Even among the Caddo, however, there had been a sharp reduction in population, with the total estimated at about fifteen hundred. William Dunbar thought there were no more than two hundred warriors in 1804, when his expedition explored the Ouachita River for the United States government, and the small size of the Caddo villages made them the object of frequent attacks by the Osage. The Caddo had experienced relatively little white contact up to this point, although occasionally they were visited by French and Spanish traders. The spread of the American frontier changed their situation. In 1818 the United States goverment established a factory, or trading post, for the Caddo at the junture of the Red and Sulphur rivers in extreme southwest Arkansas, near present-day Doddridge. During the early part of the nineteenth century the Caddo drifted slowly westward into Texas and Oklahoma.

15. INDIAN SETTLEMENTS IN THE EIGHTEENTH CENTURY, A.D.

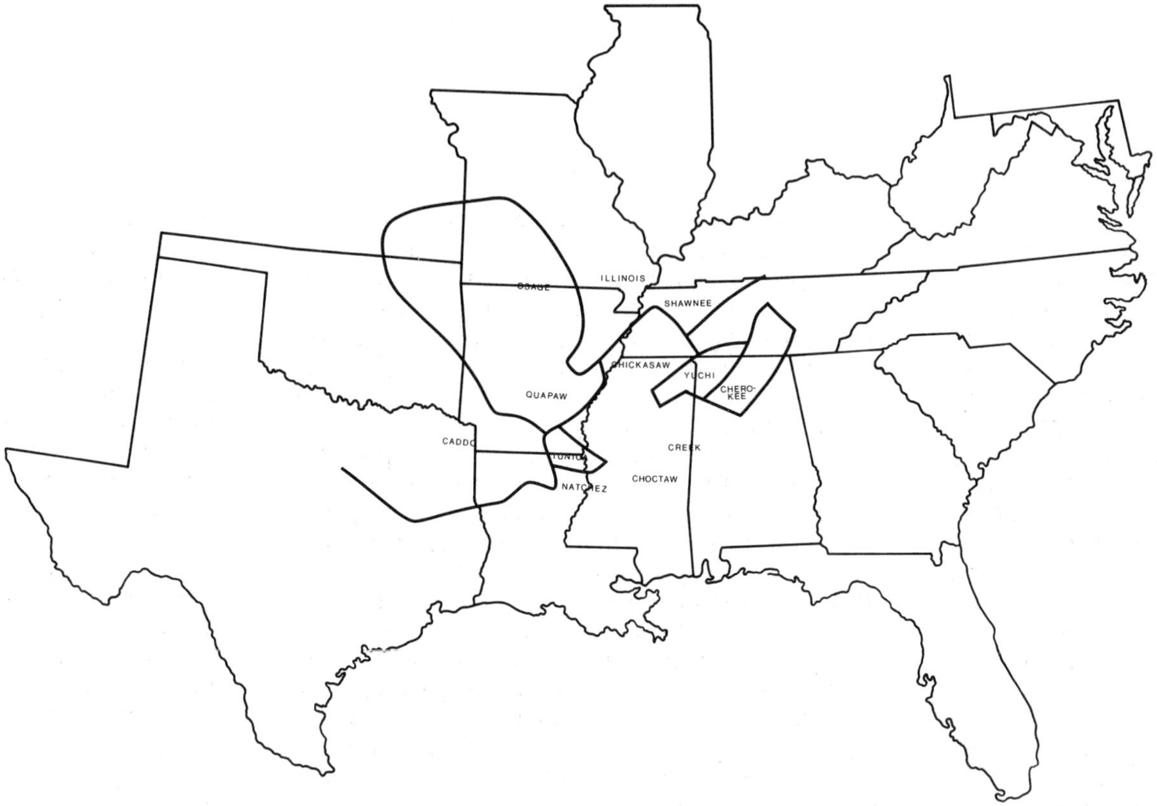

16. INDIAN TRIBES IN PROXIMITY TO ARKANSAS, 1700

16. INDIAN TRIBES IN PROXIMITY TO ARKANSAS, 1700

Before 1800, Arkansas was a crossroads of Indian culture. The Quapaw were the only indigenous Arkansas tribe, and during much of their history they were influenced by the tribes around them, either through commerce or war. These contacts increased after 1689, when the English and French promoted war among the Indians as a part of their own continuous conflict.

The most important Indian influences in Arkansas came from tribes already discussed—the Osage and the Caddo, located in contiguous territories. From their settlements on the upper Osage River and the south bank of the Missouri, the Osage raided in the Arkansas area throughout the historical period. The Caddo comprised three lesser confederations: the Kadohadacho, living in southwest Arkansas, and the Hasinai and the Natchitoches in modern-day northeast Texas and northwest Louisiana. The excellent Caddo pottery was widely traded among the Indians of the Arkansas River valley.

Many powerful Indian societies were located east of the Mississippi near Arkansas, but they had little influence on the Arkansas populaton until the eighteenth century. At that time wars among themselves and against the English settlements along the Atlantic coast began to dislocate some of these tribes and push them into Arkansas.

The major contacts from up the Mississippi River were made by the Michigamea, a subdivision of the Illinois Indians. As early as 1673 the Illinois were living along the Mississippi River somewhere between present-day Mississippi and Phillips counties. By 1718 the Michigamea had settled among the Quapaw and perhaps were allied with them in a war with the Chickasaw. Sometime after this they were defeated in a war with either the Quapaw or the Chickasaw and returned to the area of present-day Illinois.

The Tunica were located on the southeast, concentrated during the seventeenth century along the Yazoo River and in northeastern Louisiana. They were allies of the French, and ultimately they were driven from the Mississippi by a war, or a threatened war, with the Chickasaw, who were allies of the English. The Tunica moved downriver and up the Red River to Marksville Prairie in central Louisiana.

Another eastern ally of the French were the Yuchi, who lived along the Tennessee River from Muscle Shoals to the headwaters. An eastern branch of the tribe ranged widely in eastern Georgia and Florida. At war with the English and the Cherokee during the eighteenth century, the Yuchi were ultimately absorbed by the Shawnee and the Creek and moved through Arkansas with those tribes to Indian Territory in the nineteenth century.

The Choctaw lived in the southern Mississippi and Alabama area. They were allies of the French, who considered them a major barrier between their Louisiana settlements and the expanding English colonies along the Atlantic coast. Both European powers courted the Choctaw in the eighteenth century until 1750, when the tribe split over the issue and the French interests won. In 1750 the Choctaw fought the Creek and the Chickasaw. By 1819 some of the Choctaws had moved into Arkansas and were living with the Quapaw. In 1825 they would leave Arkansas for Indian Territory, and the treaty that they signed established the present Arkansas-Oklahoma border south of Fort Smith.

The Natchez were also on the eastern bank of the Mississippi, with their towns near present-day Natchez, Mississippi. Although they were usually under French influence, they also frequently fought their sometime allies, and in 1718, 1722, 1723, and 1730 the French mounted military expeditions against the Natchez. After the 1730 war the Natchez scattered, some moving west of the Mississippi and others joining the Cherokee on the east.

The Chickasaw in northern Mississippi were the chief ally of the English in the Mississippi River valley in the eighteenth century. Encouraged by the English, they warred upon neighboring tribes throughout the 1700's. Their military successes became a major force dislocating many of the Indians in the Louisiana country who allied themselves first with the French and after 1762 with Spain.

The Creek Confederation comprised several tribes, the most important of which was the Muskogee. In about 1700 elements of this group were located on the Coosa and Tallapoosa rivers in modern-day Alabama and the Flint River in Georgia. The Yamasee War of 1715 forced the Creek away from the Atlantic coast and pressured them westward. They would remain, however, in their old areas until they were removed through Arkansas to the Indian Territory after 1830.

The Cherokee were the largest tribe in the southeastern region of North America in 1700. At that time they were located primarily at the headwaters of the Savannah River in what is now western North Carolina and eastern Tennessee. During most of the eighteenth century the Cherokee traded with the English and supported the Crown in frontier wars. They were traditional enemies of the Shawnee, who occupied the Cumberland River valley of Tennessee, and with their Chickasaw allies the Cherokee expelled the Shawnee in about 1712. At the time of the American Revolution they opposed the revolutionary colonists—a politically disastrous move, considering the outcome of that rebellion.

The Shawnee ranged from southeastern Georgia to Ohio in the seventeenth century. As French allies, they were the object of frequent attacks by the Cherokee and the Chickasaw. These wars pushed them into the area of the Talladega mountains and Tallapoosa River in Alabama. Both the Cherokee and the Shawnee would travel through Arkansas during the Indian removals of the 1820's and 1830's.

16. INDIAN TRIBES IN PROXIMITY TO ARKANSAS, 1700

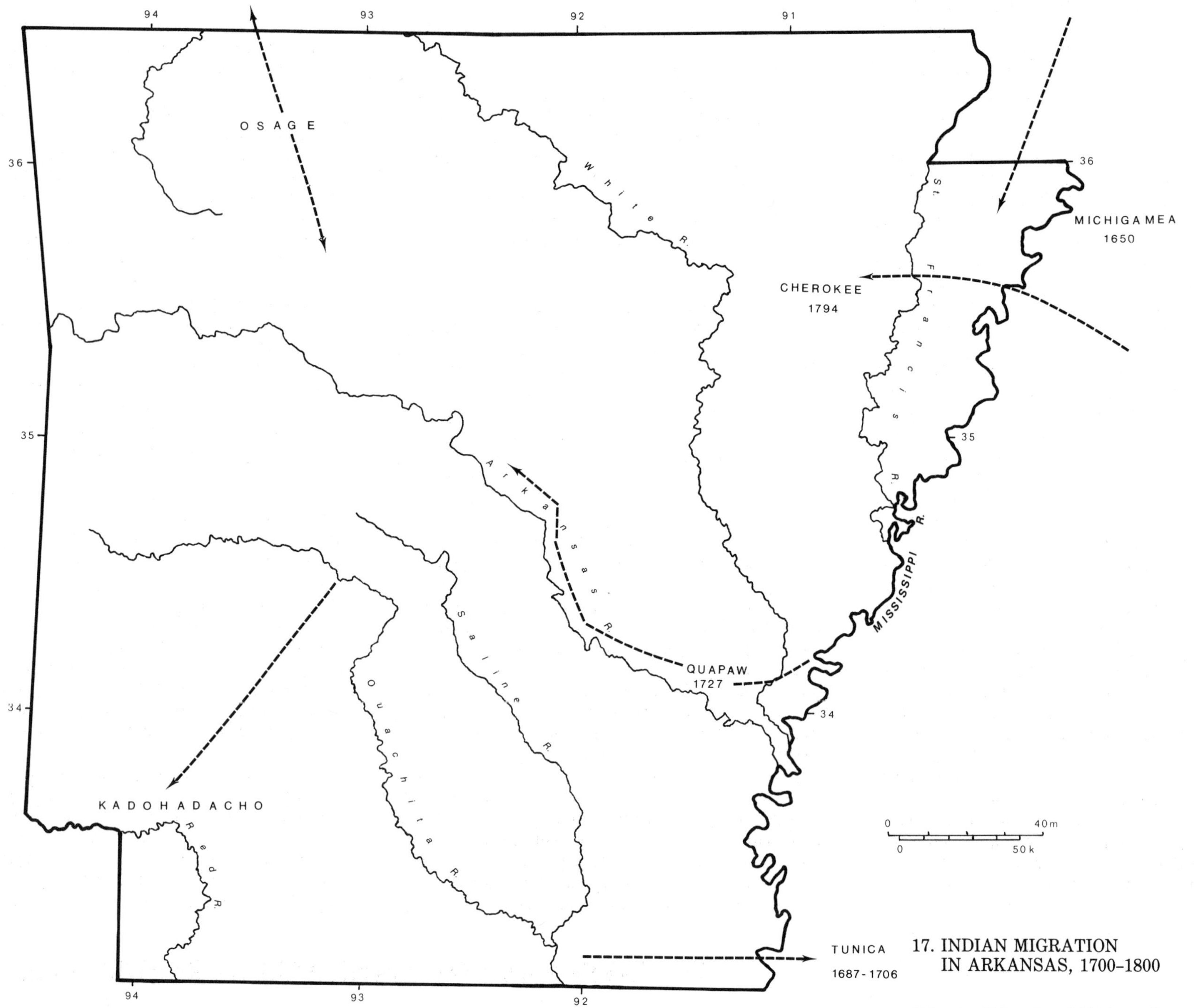

17. INDIAN MIGRATION
IN ARKANSAS, 1700–1800

17. INDIAN MIGRATION IN ARKANSAS, 1700–1800

Several major Indian migrations took nearby tribes into and through the Arkansas area. The first of these movements in the historical period took place between 1700 and 1800. The Michigamea had been the first tribe to move into the area during the historical era, arriving about 1650. Marquette found the Michigamea in northeastern Arkansas in 1673. About 1700 they were involved in wars with the Quapaw and the Chickasaw. Whether because of the wars or for other reasons, they subsequently abandoned Arkansas and joined the Kaskaskia in Illinois.

The Quapaw, originally concentrated around the mouth of the Arkansas River, moved up the river around 1727. The causes of this movement are not clear, although the Quapaw were involved in wars with neighboring tribes throughout the early eighteenth century. Their movement may have been the result of these conflicts. In 1748 or 1749 they were defeated by the Chickasaw in the vicinity of Arkansas Post after they had allied themselves with the French.

Sometime between 1687 and 1706 the Tunica left southeastern Arkansas and moved downstream to the mouth of Red River. This movement resulted from a conflict with the Chickasaw, who were working for the English during this period. Sometime after 1784 the Tunica left the Mississippi River area completely and migrated up the Red River to Marksville Prairie, in what would become north-central Louisiana. Meanwhile, in the southwest the Kadohadacho concentrated along the Red River and moved into present-day Louisiana, Texas, and Oklahoma.

By the end of the eighteenth century an additional force pressuring the Indians who had traditionally lived in Arkansas was the migration of other tribes into the area from east of the Mississippi River. In 1794 the first Cherokees moved into Arkansas. They were a band, led by The Bowl, who were fleeing Tennessee because of their role in the Muscle Shoals massacre. They settled in the swamps along the Saint Francis River in northeastern Arkansas and the Missouri bootheel. While they were later acquitted of responsibility in the massacre, these Cherokee remained on the lands that they had occupied. By the end of the eighteenth century other Cherokees, from the Lower Towns east of the Mississippi River, had begun to move into Arkansas and settle on the Arkansas River at the foot of the Ozark highlands, near present-day Russellville. Around 1800 the Cherokee in the Saint Francis River area moved to join those of their tribe along the Arkansas. This movement put the Cherokee into a direct confrontation with the Osage for control of northwestern Arkansas.

17. INDIAN MIGRATION
IN ARKANSAS, 1700–1800

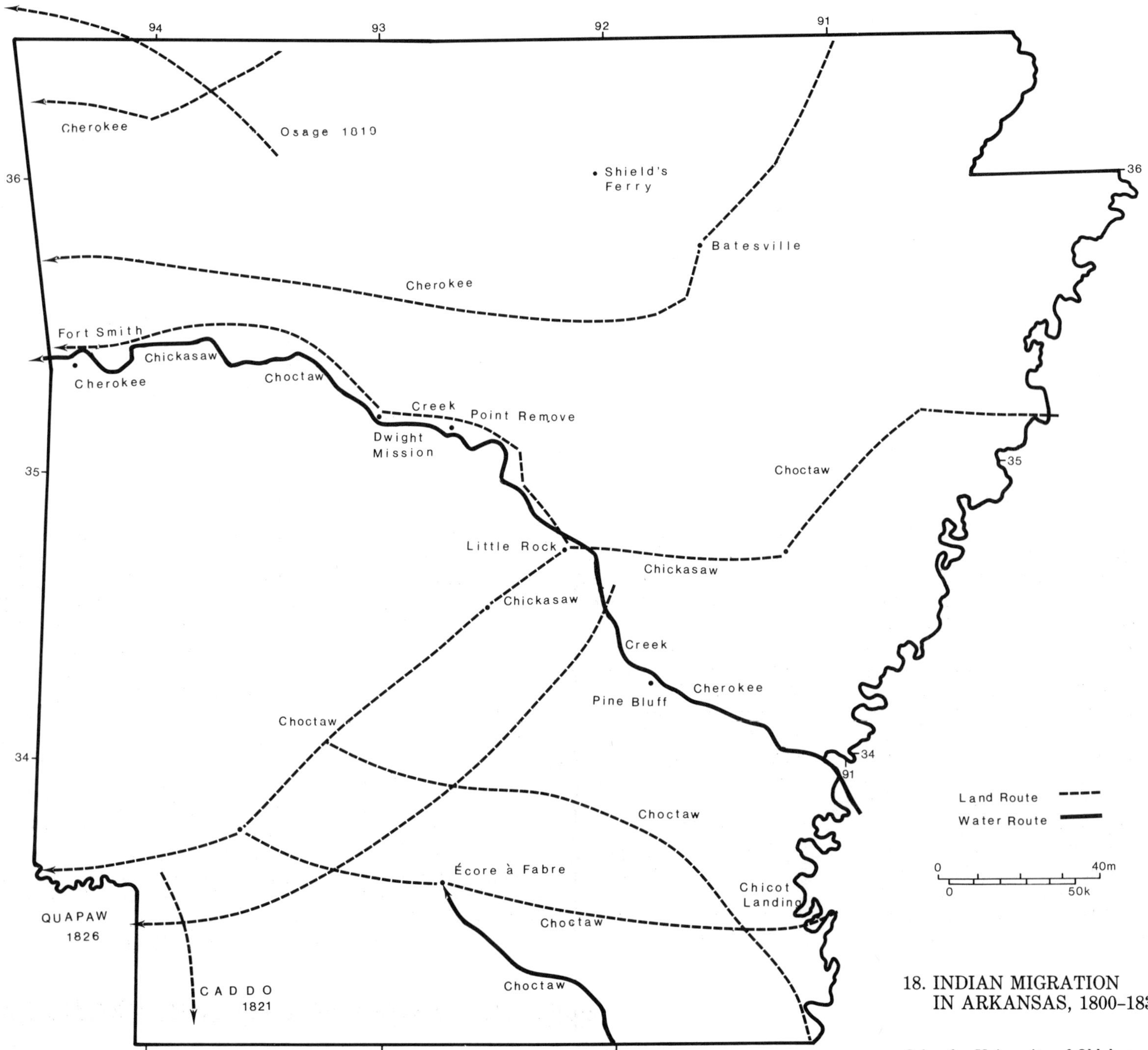

18. INDIAN MIGRATION IN ARKANSAS, 1800–1836

18. INDIAN MIGRATION IN ARKANSAS, 1800–1836

After 1800 white settlement began to place further pressure on those Indian tribes that remained in Arkansas. The Stephen Long expedition of 1820 found that most of the Quapaw had moved from the lower Arkansas River. Long reported that most of that tribe were located on the south bank of the Arkansas. Probably they were in the vicinity of modern-day Pine Bluff. They had been joined at this time by the Choctaw and remnants of other tribes from east of the Mississippi. In 1824, as a result of the Quapaw treaty of that year, the tribe lost their 1818 reservation between the Arkansas and the Saline, and subsequently they abandoned the central part of modern-day Arkansas to join the Caddo along the Red River in the southwest.

At the same time that local Indian populations were being removed, Arkansas became a point of relocation and a major route travelled by the large numbers of Indians being removed from the states east of the Mississippi to the Indian Territory, which was being established in the western part of Arkansas.

The first of these migrations involved the Cherokee. In 1817 the first Cherokee treaty ceded land east of the Mississippi for an area in north-central Arkansas between the Arkansas and White rivers, west of a line from Point Remove on the Arkansas to Shield's Ferry on the White. About thirty-five hundred Cherokee moved during this period, mainly after 1820. They established the Cherokee Nation West, in villages on their new lands, where they farmed and raised livestock. Dwight Mission, opposite Dardanelle, was the site of their principal village. Greatly influenced by their contact with white culture in the East, the Cherokee had appropriated many Anglo-English institutions, including slavery, which they brought with them to Arkansas. Among the Cherokee residents in Arkansas was Sequoyah, who apparently worked on his Cherokee alphabet while in the region.

Between 1831 and 1839, Arkansas was along the route taken by thousands of Indians removed from other southern states to the Indian Territory by the federal government. The first of these groups were the Choctaw, who were removed from Mississippi in three parties, beginning in October, 1831, and continuing into the winter of 1833. The first party was moved in four groups. One marched directly overland from Chicot Landing to Écore à Fabre near modern-day Camden, then on to Fort Towson in Indian Territory. Other Choctaw were moved up the Ouachita River, then marched overland from Écore à Fabre. Two other parties were taken up the Arkansas River to Little Rock, then marched overland to Fort Smith. These routes became the roadways for later migrants. By 1833 some 12,500 Choctaws had been removed from Mississippi to the west.

The removals continued as the Creek, the Chickasaw, and the first Seminoles were taken to the Indian Territory in 1832. The majority of these migrants traveled by boat up the Arkansas River to Fort Smith, but the usual Arkansas transportation problems caused by low water forced many to be transported overland. The Seminoles followed the Choctaw route up the Ouachita as well.

In 1828 the federal government dissolved the Cherokee Nation West and moved this group of Arkansas settlers into the Indian Territory. In 1838 the government began to remove all of the Cherokees who remained in the southeastern United States to the Indian Territory as well. The routes that the latter group traveled became known as the Trail of Tears, because of the hardships suffered by the Indians. There was not just one route, however. The Cherokee migration routes included one out of southeastern Missouri and across the mountains to Fort Gibson in Indian Territory; another from Memphis, Tennessee, to Little Rock and up the Arkansas; another from Cape Girardeau, Missouri, on the Mississippi, to Batesville and across the mountains. The Indian removal through Arkansas caused great suffering for these people. Because they lacked adequate food and clothing during the hard winter months, their route was marked by the graves of their dead. Something like eighteen thousand Cherokee moved during this period, and perhaps as many as four thousand died along the way.

18. INDIAN MIGRATION
IN ARKANSAS, 1800–1836

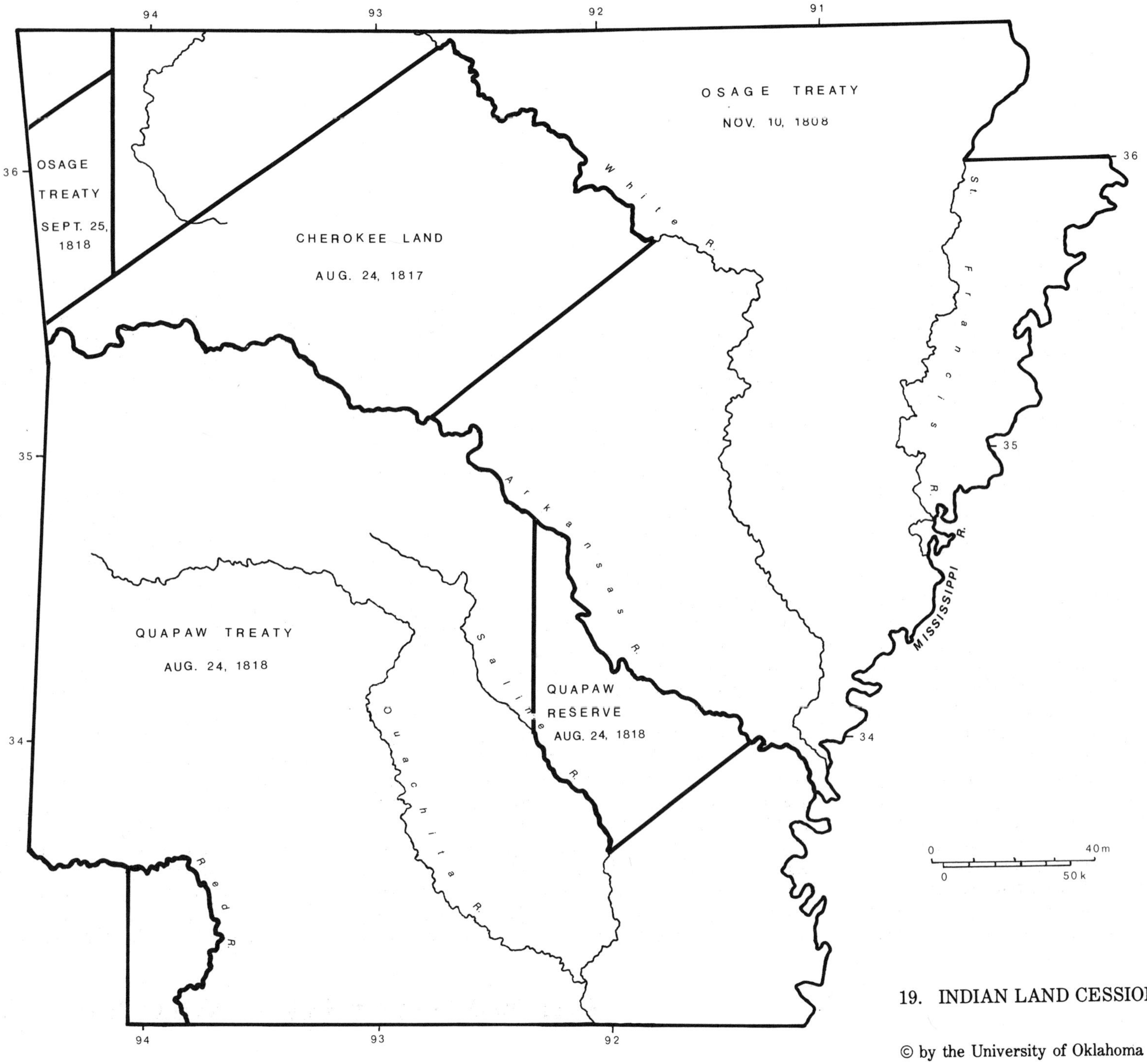

19. INDIAN LAND CESSIONS, 1803–1818

19. INDIAN LAND CESSIONS IN ARKANSAS, 1803-1818

After the Louisiana Purchase in 1803 the westward movement of Americans put increasing pressure on the Indians to give up their land. The organization of the Louisiana Territory for settlement resulted in numerous treaties in which Indians surrendered their claims to land in what would be Arkansas. The first of these agreements was signed on September 25, 1808, when Pierre Chouteau, representing Louisiana Territorial Governor Meriwether Lewis, signed a treaty with the Osage at Fort Clark on the Missouri River, in modern-day North Dakota. This treaty ceded all Osage land east of a line—called both Brown's line for its surveyor, and the Fire Prairie Line for its point of origin—from Fort Clark to Frog Bayou on the Arkansas River, in northwest Arkansas, and north of the Arkansas to the Mississippi. As a result of this treaty the Osage surrendered over fourteen million acres in what is now Arkansas. A supplementary treaty was signed on November 10, 1808, after various chiefs complained that they had not intended to surrender land but only hunting privileges. The Osage retained the land west of the boundary established in the first treaty in northwest Arkansas, but they gave up the greater portion of their claims. In return for this cession, the Osage received a garrison at Fort Clark to protect "orderly Indians," a government trading post, a blacksmith, a mill, an annual government payment of cash and merchandise, and government settlement of claims by white settlers against the Osage.

Another Osage treaty was negotiated by William Clark, governor of the Missouri Territory, at Saint Louis on September 25, 1818. This cession was made in an effort to settle numerous claims against the Osage for damages done by raiding parties in the decade since the first Osage treaty. The government had taken responsibility for these claims, paying settlers out of the annual Osage settlement; but by 1818 the payments for damages were enough that the Indians would have received no money or supplies, and would not for several years. This 1818 treaty ceded that portion of the land left to the Osage from the boundary line at Frog Bayou in Crawford County on the Arkansas, west up the Arkansas to the Verdigris River in the future Indian Territory, then up the Verdigris to the falls, and then northeast to the 1808 line at a point twenty leagues north of the Arkansas River (see Map 27). This left the Osage a small amount of land in present-day northwest Arkansas, which they did not cede until 1825. In return for the 1818 cession the United States agreed to pay for all property destroyed by the Osage after 1814. A final Osage treaty, signed on June 2, 1825, liquidated the tribe's last claims in Arkansas, surrendering any remaining lands lying in the territory of Arkansas and the state of Missouri.

At the same time that the Osage were surrendering their claims in Arkansas, the Cherokee were moving in. In 1809, President Thomas Jefferson authorized the settlement of Cherokee who desired to remove from east of the Mississippi to an area between the Arkansas and White rivers. Many migrants began this removal, but they found themselves involved in constant conflict with white immigrants over territory. To settle such disputes they asked in 1815 for the assignment of specific lands in Arkansas. A treaty signed with the tribe on August 24, 1817, granted land to them in northwest Arkansas. In turn, the Cherokee surrendered, acre for acre, an equal amount of land claimed by them east of the Mississippi. The territory given to the Cherokee in this treaty included land between the Arkansas and White rivers, west of a line between Point Remove on the Arkansas to Shield's Ferry on the White, above Batesville. No western boundary was established until 1825, when one was surveyed from Table Rock Bluff, on the Arkansas about ten miles above Fort Smith, due north for a distance of one mile and seventy chains, then northeast to the White River opposite the mouth of Little North Fork.

On August 24, 1818, the Quapaw ceded all their land claims south of the Arkansas River in return for the creation of a reservation. The new reservation's boundaries, as described in the treaty, ran from Arkansas Post in a southwest direction to the Ouachita River, from there to the Saline, and then up the Saline to a line running due north to the Arkansas River at the "little rock," then downriver to the starting point. In fact, the line hit the Saline before encountering the Ouachita, and the reservation was thus bounded on the southwest by the Saline alone. In return for this cession, the Quapaw received immediately from the United States four thousand dollars in merchandise and the promise of an annual payment of one thousand dollars in merchandise thereafter.

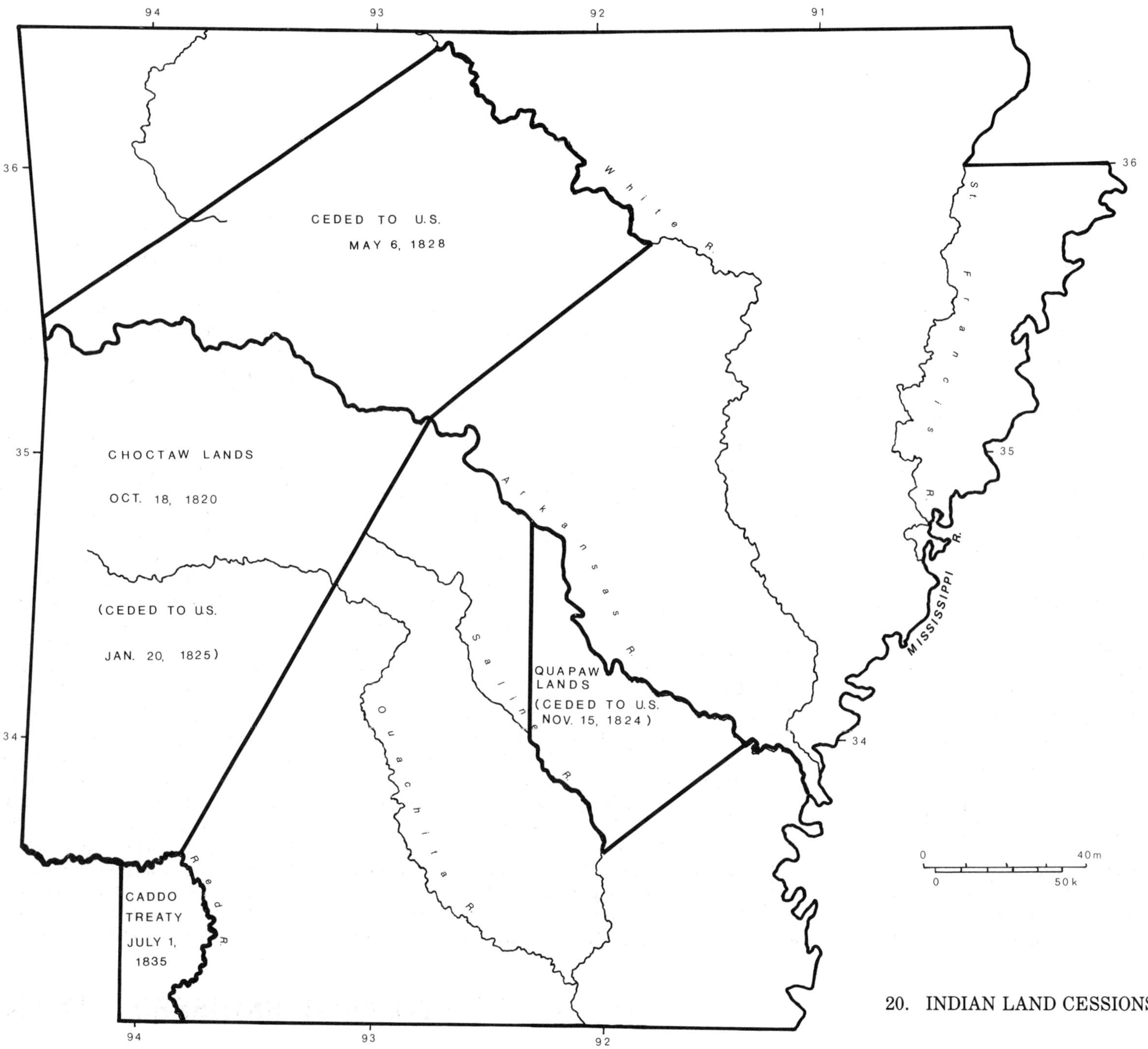

20. INDIAN LAND CESSIONS, 1818–1835

20. INDIAN LAND CESSIONS, 1818–1835

At the time that the most efforts were being made to remove Indians from Arkansas Territory, the United States government used a part of western Arkansas as a reservation for the removal of the Choctaw from east of the Mississippi. In the treaty of Doak's Stand, October 18, 1820, the Choctaw received a strip of land in western Arkansas that ran from a point on the south bank of the Arkansas River opposite the mouth of Point Remove Bayou upriver to the Canadian, then up the Canadian to the western boundary of the United States at the 100th meridian, south along the meridian to the Red River, down the Red to a point three miles below the mouth of the Little River, then on a northeast line to the point of origin. The treaty created an uproar among white settlers in the Arkansas Territory, who forced the government to revoke the treaty. When Henry D. Downs surveyed the eastern line of the cession in 1821, he found white settlers already on land west of that line and recommended that it be drawn farther west. While the goverment refused to act on this recommendation, the pressure of white settlers ultimately forced the federal authorities to adjust the Choctaw boundaries.

The second Choctaw treaty, signed on January 20, 1825, was an effort to resolve the problems created by the 1820 treaty. In this treaty the Choctaw surrendered the land that they had received in 1820 that was east of a line beginning on the Arkansas River "one hundred paces" east of Fort Smith and running due south to the Red River (see Map 28). In return for this settlement, the Choctaw received an annual payment from the United States goverment of six thousand dollars. That the line was not run due south, but somewhat southwest, was the source of another controversy between the Choctaw Nation and Arkansas. This dispute was settled in 1877 with the United States compensating the Choctaw for the surveying error.

White population growth in Arkansas Territory after 1819 placed increasing pressure on the Quapaw to surrender the reservation that they had received in 1818. They gave up all their claims to Arkansas land in a treaty negotiated by Robert Crittenden on November 15, 1824. The treaty provided for the resettlement of the Quapaw among the Caddo in the southwestern part of the state. In turn, the four principal chiefs of the Quapaw received five hundred dollars to repay them for their losses, and the tribe received four thousand dollars in goods and one thousand dollars each year for eleven years. Many members of the tribe were reluctant to surrender their reservation on the Arkansas, and their opposition proved to be justified. They were not welcomed by the Caddo, and many returned to their old territories. Finally, in a treaty made in 1833, they were resettled in northeastern Indian Territory.

The Cherokee, who had first received land in Arkansas in 1817, confronted the same pressure to move that was felt by other tribes. Despite opposition within the tribe, the Cherokee chiefs negotiated a new treaty on May 6, 1828, in which they surrendered their land between the Arkansas and White rivers in return for 7 million acres in the Indian Terrritory. The treaty established a western boundary for Arkansas Territory running from the north bank of the Arkansas, opposite the Choctaw line of 1825, then on a direct line to the southwest corner of Missouri (see Map 27). All white inhabitants west of the Cherokee Line were required to move east. The survey of this line was completed on November 19, 1828.

The last Indian land claims in the state were given up in the Caddo treaty of July 1, 1835, in which the Caddo surrendered all of their claims on lands in Arkansas and Louisiana. Specifically, these lands were between the Red River and the Sabine River (now in northeast Texas). The territory given up in Arkansas lay within the boundaries of present-day Miller County.

PART THREE
EUROPEAN AND AMERICAN EXPLORATION

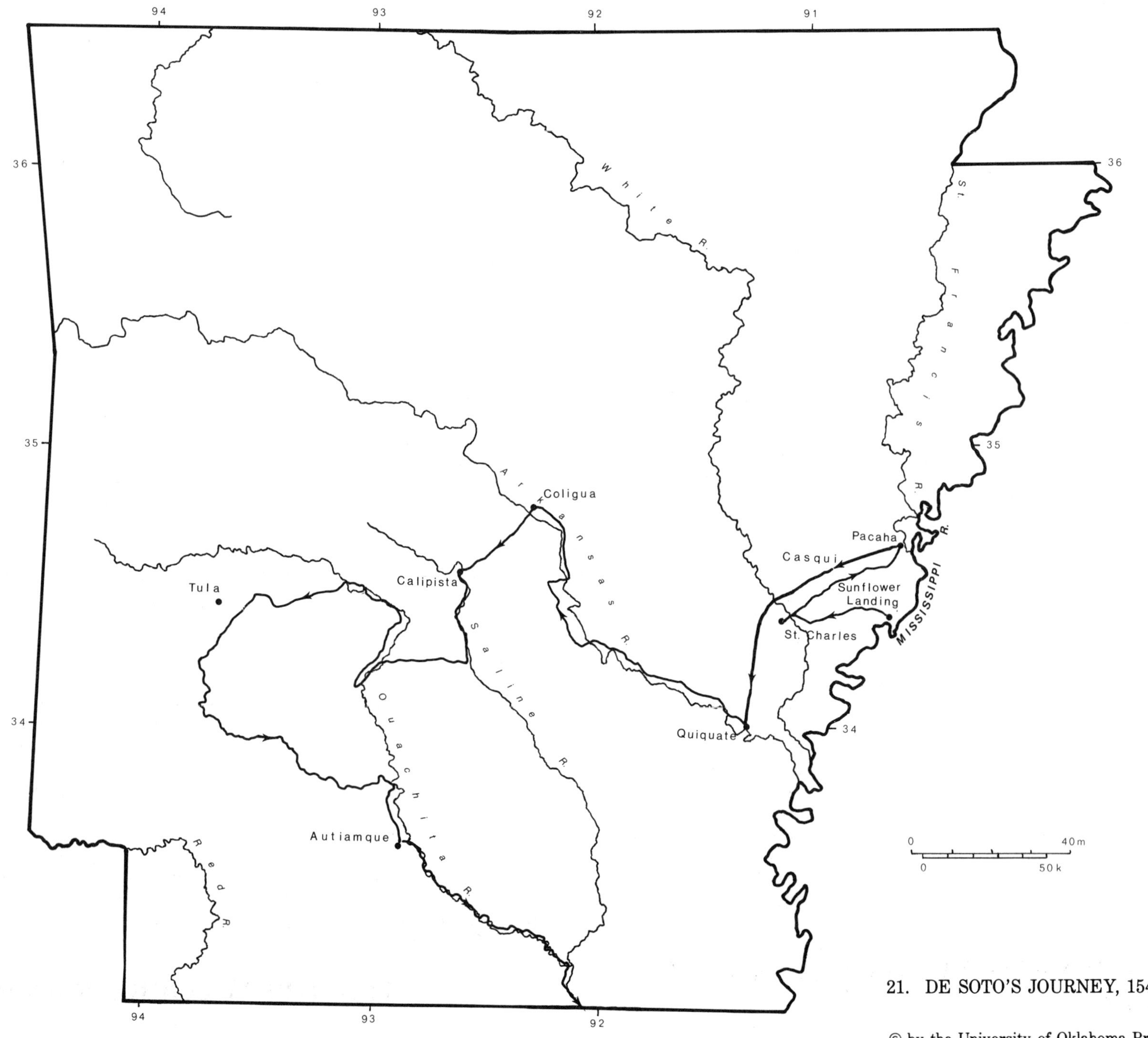

21. DE SOTO'S JOURNEY, 1541–1542

21. DE SOTO'S JOURNEY, 1541–1542

The first European to enter Arkansas was the Spanish explorer Hernando de Soto. De Soto came to Arkansas as part of his commission to explore Florida, a trip that took him through the southeastern United States and brought him to the Mississippi River in search of treasures. His travel in Arkansas has been the topic of much debate, and it continues to be discussed by historians. The United States de Soto commemorative commission, appointed in 1939 to commemorate the four-hundredth anniversary of the expedition, defined the route presented here, although this is not the only one accepted by scholars.

The key problem of any analysis of de Soto's route is defining where he came into Arkansas. Generally it is agreed that he crossed the Mississippi somewhere south of modern-day Helena, Arkansas, in June of 1541. The de Soto commission placed that entry point at Sunflower Landing, twenty miles below Helena. The first town that de Soto visited was Aquixo in a province of the same name.

The expedition went west to the White River near modern-day Saint Charles, then north to the town of Casqui, which was part of a group of settlements located either on Big Creek or the L'Anguille River. De Soto then moved north to another town, called Pacaha, near the mouth of the Saint Francis. From this point an expedition party was sent farther to the northwest and went as far as the vicinity of modern-day Wheatley, Arkansas, between the Saint Francis and White rivers. The Spaniards then turned south from Pacaha and returned to Casqui, then marched back to the White River. They moved down the west bank of the White until they arrived at a town called Quiquate in the delta between the White and Arkansas rivers, probably about seven miles east of the future Arkansas Post.

Having explored the region along the White and Mississippi, de Soto's expedition turned to the northwest and moved up the Arkansas, crossing to the south bank above Pine Bluff, then recrossing near Woodson. They came upon the town of Coligua on the north bank at Great Rock, near the site of present-day Little Rock. From Coligua they moved southwest past modern-day Benton, then followed the Saline for some distance before cutting to the Ouachita River, which they hit either near modern-day Friendship or near Donaldson. In this area the expedition spent about a month and explored the vicinity. They moved up the Ouachita to Tanico, near Hot Springs, then up the valley of the Caddo River to Caddo Gap, where they found a powerful group of Indians, called the Tula, who were probably part of the Caddo Confederacy.

The expedition's course out of Tula remains the least-clear part of their march. Even the de Soto commission failed to reach a conclusion about the route, and it offered two alternatives. De Soto possibly kept near the Caddo River until he reached the site of modern-day Arkadelphia, then crossed to the east bank of the Ouachita, from which point he headed south. A more probable route crossed the headwaters of Antoine Creek and went down the Little Missouri, then to the Ouachita. Whatever direction the expedition took, in November, 1541, it arrived at a major crossroads near Camden, which the Indians called Autiamque, Utimque, or Calion. At this place they spent the winter of 1541–42.

De Soto left Autiamque on March 6, 1542, with three hundred men and marched down the Ouachita River. The party followed this stream to the Red River, then to the Mississippi. On May 21, 1542, de Soto died of unknown causes. The party continued, ultimately arriving back in Spanish settlements in Mexico on September 10, 1543.

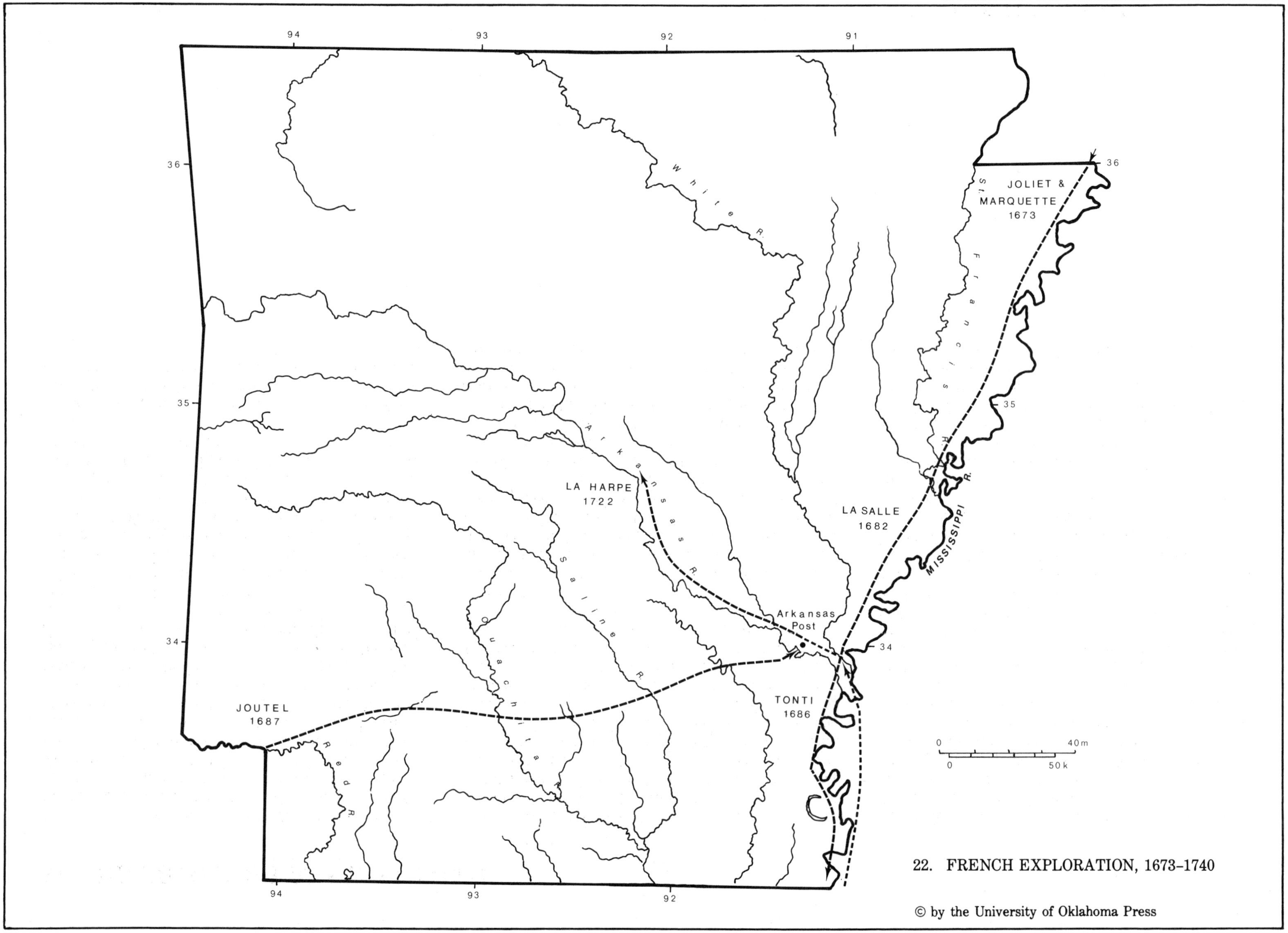

22. FRENCH EXPLORATION, 1673–1740

22. FRENCH EXPLORATION, 1673–1740

Father Jacques Marquette and Louis Joliet were the first of many French explorers who visited Arkansas. Marquette, a Jesuit missionary, received a French commission in 1673 to search for a waterway to the Pacific. Joliet, a fur trader and explorer, was sent by the French officials in Canada to accompany Marquette. Their expedition left Fort Michilimackinac on Lake Michigan on May 13, 1673. They moved by the Wisconsin River to the Mississippi River and explored down the Mississippi to the mouth of the Arkansas River. In June, 1673, the party put ashore at an Indian village called Arkansea. At this village they determined that the Mississippi flowed into the Gulf of Mexico and, desiring not to risk capture by hostile Indians or the Spanish in the south, they turned back toward Canada.

The next French travellers arrived in Arkansas in 1682, when Robert Cavelier, sieur de La Salle, came down the Mississippi exploring French claims in the area. On March 12 the French arrived at the Indian village of Kappa on the Arkansas River approximately twenty miles north of its mouth (see Map 14), which was probably the same village known to Marquette and Joliet as Arkansea. Later the expedition explored the Arkansas, moving almost thirty miles upstream. Along the way they passed other villages called Tongigua, Tourima, and Osotouy. These villages were inhabited by a people variously called the Kapaha and the Akancas. Before leaving the area, on April 9, 1682, La Salle erected a cross on high ground and claimed the region for France.

La Salle was indirectly responsible for the third French entry into the region. In 1684 he led an expedition that attempted to discover the entrance to the Mississippi River from the Gulf of Mexico. His party failed and became lost, and in 1686, French authorities ordered Henri de Tonti, a former lieutenant of La Salle who was at Fort Saint Louis on the Mississippi, to search for the lost Frenchman. Tonti was unable to find La Salle at the mouth of the Mississippi and returned up the river. On the way back, however, he stopped at the Arkansas villages, where some of his men requested permission to remain. Tonti allowed them to stay, in part hoping to establish contact with La Salle, and they built a settlement near the Indian villages along the river. This location, which is on the edge of Little Prairie and is now called the Menard site, became Arkansas Post, the first permanent white settlement in the Arkansas area.

The expedition of La Salle was at this time marching northeast in what would become Texas. Somewhere along the Trinity River, La Salle's men mutinied and murdered him on March 19, 1687. While some men deserted the expedition at this time, others continued their march to the northeast under Henri Joutel, hoping to find the Mississippi and work their way north towards Canada. Joutel's party crossed the southern part of Arkansas and arrived at Arkansas Post on July 24, 1687.

In 1722, Bénard de La Harpe explored the Arkansas and Red rivers. This was the first official exploration along the Arkansas by the French. La Harpe's trip was spurred by rumors among settlers in the John Law colony in the vicinity of Arkansas Post that a large emerald rock was located on the river. La Harpe attempted to find the emerald. The expedition organized at Arkansas Post and moved upriver, camping at what is known as Big Rock (near modern-day Little Rock), which they named La Rocher Française. While they did not discover the emerald, they reported that they had found a land with considerable game and ample minerals. One of the members of the expedition thought that there might also be gold in the country. La Harpe went west as far as the present-day city of Fowler, Arkansas, then turned back because his men opposed any further westward movement. La Harpe mapped the river, although the map has not survived.

Two more Frenchmen, the Mallet brothers, travelled with their party through Arkansas in 1740. Moving from Santa Fe towards the Platte River, they came across the Arkansas River near Dodge City, Kansas. They followed it in hopes that it would join with the Platte or the Missouri. They soon found that it did not, but followed its course to the Mississippi. When they reached the Mississippi, the expedition turned southward for New Orleans.

The French left a lasting impact on the landscape of Arkansas. As French explorers and fur trappers moved through the region they named many of its natural features. The unlabelled streams on this map all have French place-names and indicate the extent of French activity in the area.

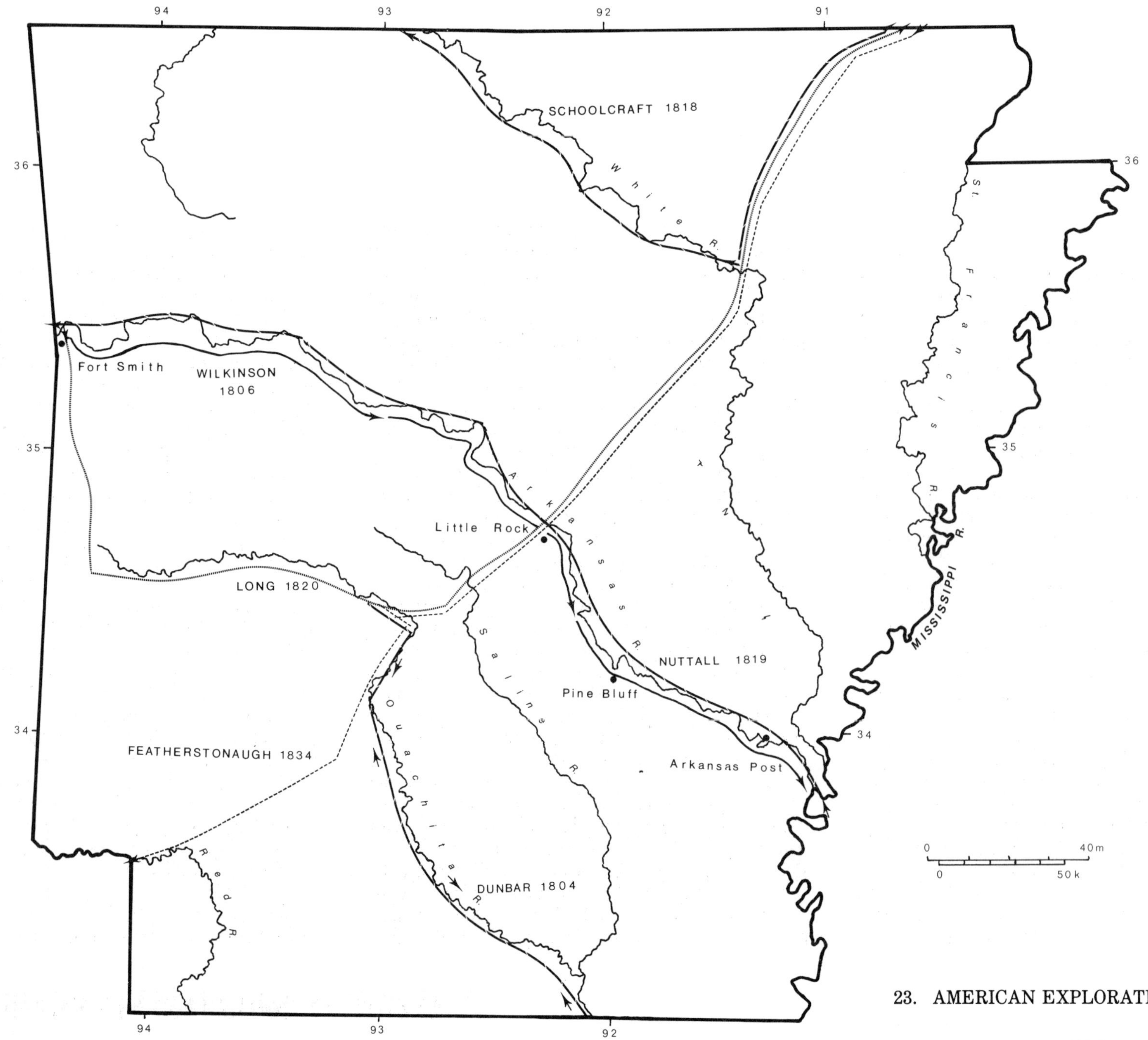

23. AMERICAN EXPLORATION, 1803–1834

23. AMERICAN EXPLORATION, 1803–1834

Arkansas continued to be an object of exploration after the American purchase of the area in 1803. The first official expedition into the area came in 1804 and 1805 when President Thomas Jefferson commissioned William Dunbar to explore the Ouachita River. Dunbar and a companion, Dr. George Hunter, entered what would be Arkansas on November 16, 1804. They found the Ouachita difficult to travel, but were able to move up the river as far as the Hot Springs, the chief object of their expedition. They found a country filled with wild game and discovered a few whites trapping and trading with the Indians. The Quapaw, or Arkansas Indians, remained along the Arkansas River in three villages, and Dunbar reported the tribe to consist of 260 warriors. Other than the Quapaw, Dunbar encountered few Indians.

The next major expedition to chart parts of Arkansas came in 1806, when Lieutenant James B. Wilkinson was detached from an expedition to the Rocky Mountains led by Zebulon M. Pike. On the upper waters of the Arkansas, in present-day Kansas, Pike decided to send Wilkinson down the river. Wilkinson and three other men undertook the journey in two canoes. Wilkinson mapped the river, and he observed settlements of the Quapaw and the Choctaw, but met few other human inhabitants in the area. The four men reached Arkansas Post on January 9, 1807. Unfortunately, Wilkinson was ill when he reached the settlement and, for that reason, did not write much about the people he encountered there.

In 1817, Major Stephen H. Long led a party up the Arkansas River on his way to establish an army post at the point of the Poteau River's confluence with the Arkansas. Long named this place Belle Point, and the post he established was called Fort Smith. In his report of the expedition Long noted the presence of numerous Indian communities in the lowlands, including parties of eastern Indians, such as the Delaware and the Shawnee, who had moved into the west in the face of white expansion. In 1820, Long again entered Arkansas on a trip down the Arkansas River. At Fort Smith he left the river and marched across northern Arkansas to Cape Girardeau, Missouri.

Henry R. Schoolcraft was a writer, ethnologist, and geologist who explored Arkansas and spent most of the winter of 1818–19 in the northwest portion of the territory. He noted local Indians and the Ozark Mountains. The few white families that he encountered were engaged primarily in hunting and lived almost totally isolated from the outside world. Schoolcraft entered the area at Mammoth Spring and moved down the Spring River to the Black, then back into the Ozarks along the White. In the spring of 1819 he travelled down the White River, visited the site of Calico Rock and Poke Bayou, then turned northward along the route from Arkansas Post to Saint Louis.

Thomas Nuttall, an English naturalist, came in 1819 gathering information for a book he proposed to write on American flora. He arrived at Arkansas Post on January 22. His pupose was to explore the Arkansas River and study it's flora. Nutall explored the Arkansas River to Fort Smith. He noted that by 1819 the area was beginning to be dotted with the cabins of settlers and small settlements. Communities had emerged along the Arkansas River at Pine Bluff, in the vicinity of Little Rock, and at the mouth of Cadron Creek above Little Rock. On May 19, 1819, the botanist and a half-dozen soldiers left Fort Smith and moved west into the Indian Territory.

Among the last of the great early travel writers to visit Arkansas was George William Featherstonhaugh, who came in 1834. This English geologist and his son visited Arkansas as they travelled between Missouri and Louisiana, entering the territory from the north in what is now Randolph County. Their course took them down the Southwest Road to Little Rock and on to Fulton on the Red River. From Fulton, Featherstonhaugh retraced his path to Little Rock. From Little Rock the travellers took a steamboat down the Arkansas to Arkansas Post and then through the cutoff to the White River, where they entered the Mississippi. Featherstonhaugh journeyed through Arkansas at a time of transition. He was one of the last of the "explorers" and the first of the "excursionists."

PART FOUR
LAND SUBDIVISION AND BOUNDARY CHANGES

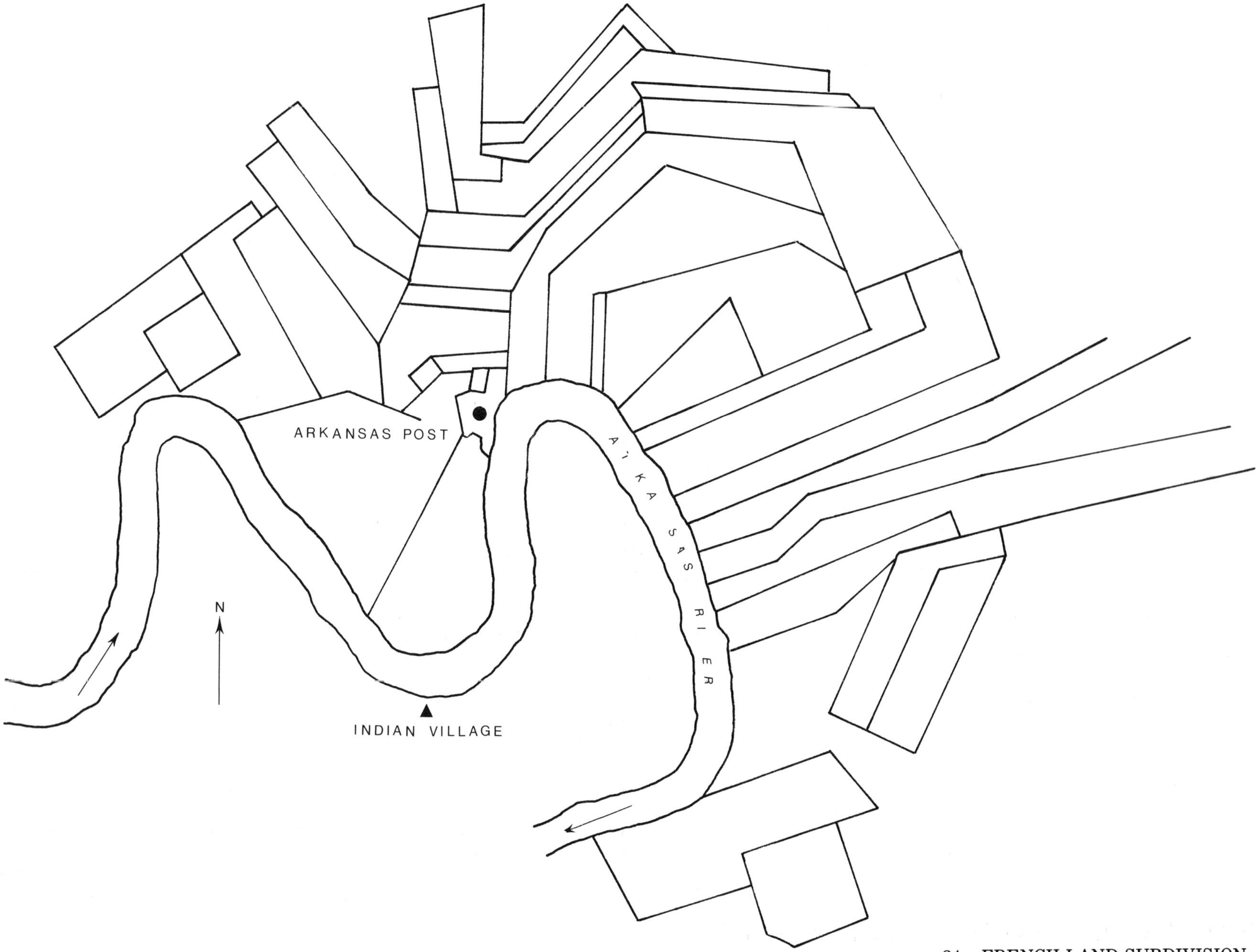

24. FRENCH LAND SUBDIVISION

24. FRENCH LAND SUBDIVISION

French claims to the land that would be Arkansas dated from April 9, 1682, when Robert Cavalier, sieur de La Salle, took possession for France of all the country watered by the Mississippi and its tributaries. For almost eighty years after that, the French did little to occupy or develop the Arkansas area.

The first grant of land made by the French was in 1682. Under authority granted to him by Louis XIV, La Salle gave a large tract of land on the Arkansas River to his friend Henri de Tonti. Tonti had grandiose plans for this land and intended to establish some sort of manor on his holding. In 1686 he left six men at the Arkansas Post, but delays in the entire French effort in the New World, including the death of La Salle in 1686, prevented the development of his plan. When Henri Joutel visited the post in 1687, he found only two of the six still there. By the end of the seventeenth century the post was abandoned completely.

The only other French grant of land in the Arkansas area was made in 1718. It was to John Law and the Western Company, which Law had organized for the purpose of establishing French settlements in Louisiana. Chartered in 1717, the company received the power to sell whatever lands might be given to it and to appoint the governmental officers of its territories. The first grant received by the company was 82,160 acres at the mouth of the Arkansas River. In return for the grant, Law agreed to bring fifteen hundred settlers into the area and to protect them. Under Law's authority Le Page du Pratz introduced some eight hundred men into Louisiana in 1718. In 1719 additional settlers and some five hundred slaves were brought to Louisiana for settlement in the Law colony.

The first colonists arrived in Arkansas in August, 1721, and located their fields and town on the north bank of the Arkansas River near Tonti's Arkansas Post. The Colony was approximately twelve miles above the river's mouth and seven miles west of the entrance of the White River cutoff into the Arkansas. Although the exact number of people who arrived at the Law colony is not known, probably no more than 250 were ever there. In the midst of efforts at starting the colony, Law went bankrupt. As a result, most of the people sent by Law from Europe never left the area around New Orleans and were settled on the Delaire grant near that city. In 1722, Bertrand Dufresne reported that forty-seven persons were at the Arkansas Post, settled in twenty cabins. They had cleared only two and a half acres of land, however. To encourage production, Dufresne actually distributed land to some of the laborers. Little could be done to maintain the colony, and it slowly deteriorated. In 1725 the French military abandoned the post that they had established there. In 1727 a visiting Jesuit reported that only thirty Frenchmen were left.

The accompanying map shows the French long-lot survey made of landholdings for the Law colonists.

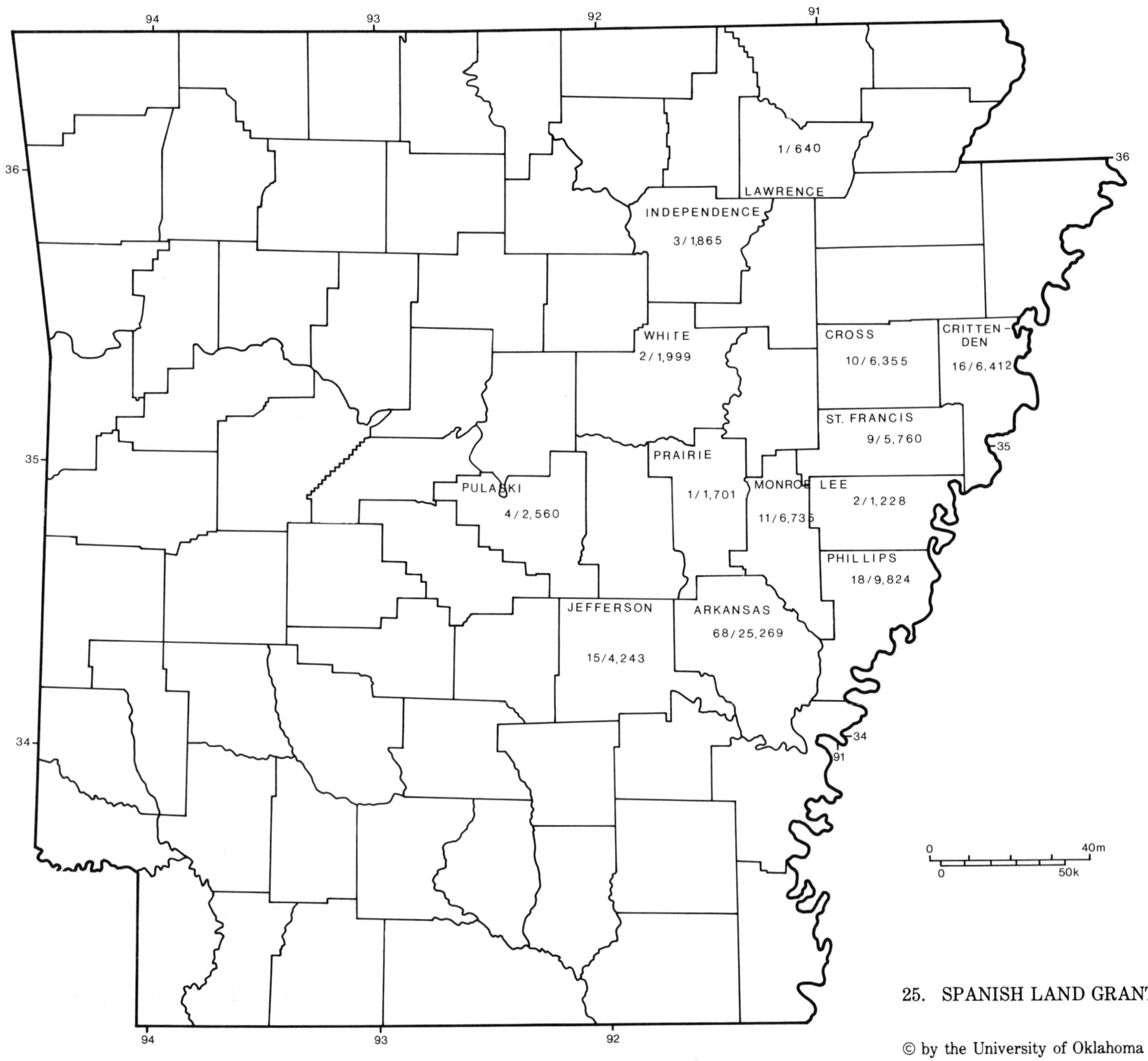

25. SPANISH LAND GRANTS IN ARKANSAS

25. SPANISH LAND GRANTS IN ARKANSAS

While Spain possessed a claim on the land that would be Arkansas as a result of de Soto's expedition in 1541 and 1542, Spain was never able to assert that claim and the land fell into French hands. Then, as the result of a secret treaty negotiated at the end of the French and Indian Wars, France ceded her claims to land west of the Mississippi. This cession was confirmed in the Treaty of Paris, signed on February 10, 1763. Spanish control was not secured until the arrival of Governor Antonio de Ulloa at New Orleans on March 5, 1766.

Francisco Luis Héctor, barón de Carondelet, Spanish governor from 1791 to 1797, witnessed with concern the expansion of the United States frontier after the American Revolution, and as a result he attempted to create a barrier to that movement through the development of Spanish settlements in Louisiana. To further that goal, he made several land grants in the area of Arkansas, the first of which was to Captain Joseph Vallière, who was given possession of the land along the White River from its mouth to its source to a depth of thirty miles on each side. Vallière never attempted any settlements, however. In 1795, Carondelet made another grant to Carlos de Villemont, commander of the Spanish garrison at Arkansas Post, for a piece of land extending six miles along the Mississippi, three miles deep, at a place about seventy-five miles below the mouth of the Arkansas River. Again, Carondelet's hopes of ensuring settlements came to naught. A third unsuccessful grant was made to the baron de Bastrop on June 20, 1797, for territory situated on the Ouachita River.

By 1797, the Americans whom Carondelet hoped to keep out were attempting to secure land grants themselves. On June 27, 1797, a group of Americans headed by Elisha Winter secured a grant of 850,000 acres in the vicinity of Arkansas Post to cultivate flax, wheat, and hemp. The land was actually occupied by the Winter family and Joseph Stillwell in 1798. They erected buildings and began cultivation on their lands. The claim to the land of these families was never made clear, and Joseph Stillwell's claim, four miles upriver from Arkansas Post, was the only one for which a proper title was obtained and a land patent received after the American takeover of the territory.

Other Americans took advantage of the Spanish officials broad power to issue land warrants. One of the most controversial of the grants was one allegedly made by the governor of Louisiana, Esteban Miró, to Jean Filhiol in 1787 for the Hot Springs in present-day Garland County. Filhiol, however, never did anything with his land, which he sold to Narcissa Bourgeat in 1803, then took back at a later date. In 1829, Grammont Filhiol attempted to reassert this claim, but a surveyor testified that Jean Filhiol had never received a regular grant, but had faked a grant and a history of land transactions to secure recognition from the United States government after Arkansas became a possession of the latter. Whatever the merits of the Filhiol claim, the courts concluded it was groundless, and they never recognized the assertions of the Filhiol family.

After the United States secured Arkansas by the Louisiana Purchase, efforts by Americans to secure legal title to Spanish lands led Congress to authorize a commission in 1805 to investigate the claims. Of the many claimants, the commission recognized the rights of only 160. The map shows the counties in which these legitimized claims were located. The first figure after the county name is the number of land grants, followed by the total acreage of the grants in the county.

SYSTEM

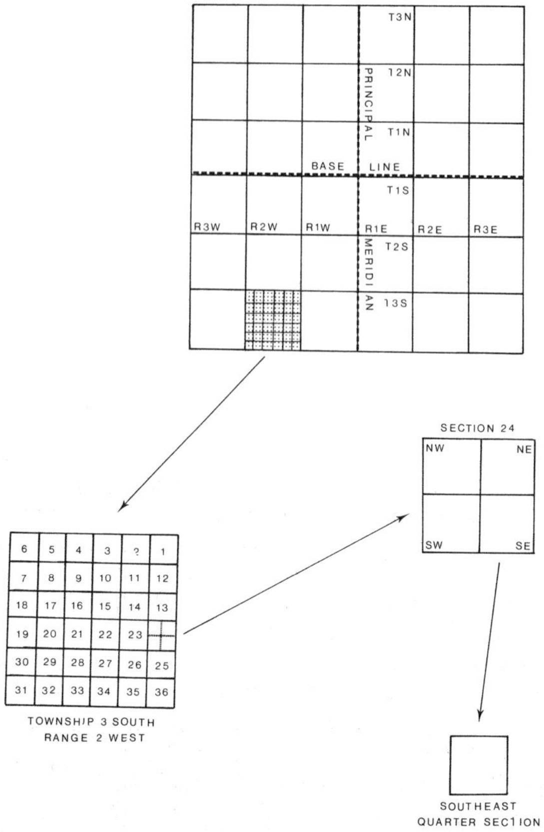

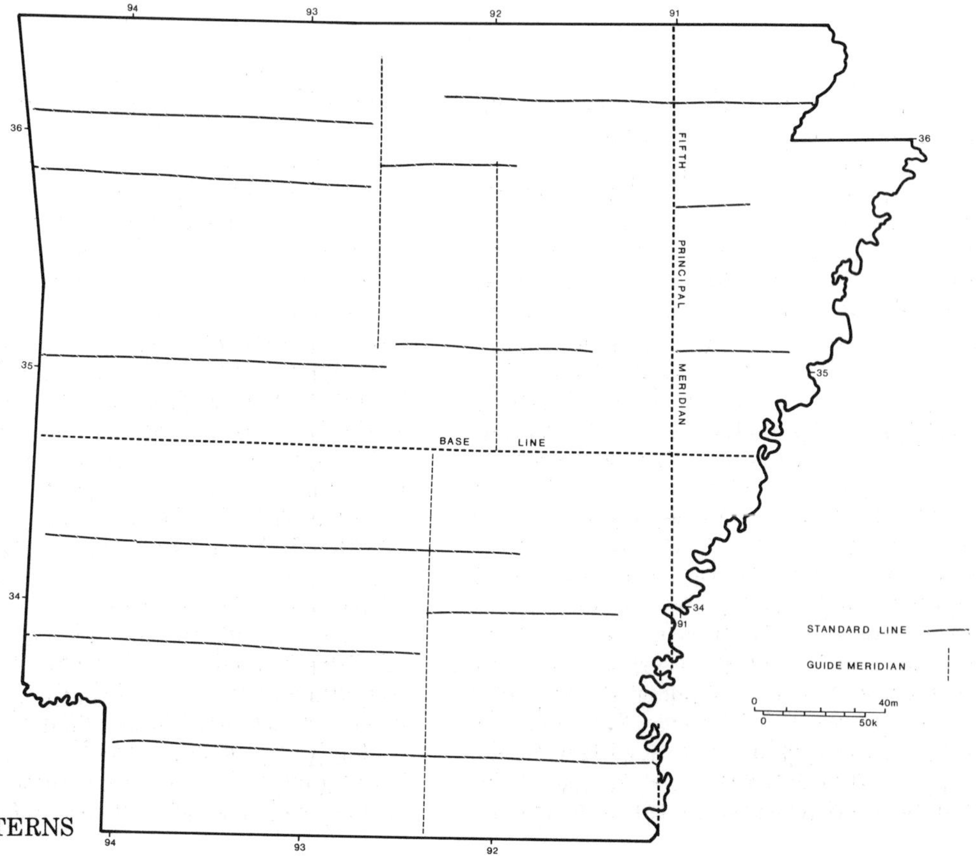

TOWNSHIP AND RANGE PATTERNS

26. TOWNSHIP AND RANGE SYSTEM

26. TOWNSHIP AND RANGE SYSTEM

Beginning with the development of the Northwest Territory in 1787, the United States government pursued a common policy for the subdivision of land for sale in all of its territories. After establishing a baseline and a principal meridian, surveyors divided land into a grid pattern by extending township lines parallel to the baseline at six-mile intervals and range lines parallel to the meridian at the same distance. The intersecting lines formed townships six miles square. Townships were identified by a system locating them with a township number north or south of the baseline and a range number east and west of the principal meridian. Each of these townships included thirty-six sections of land, each section containing 640 acres. Sections could be subdivided into quarter sections, then into 40-acre tracts.

The survey of Arkansas Territory began in 1815 when the federal government set aside two million acres of land as bounties for veterans of the War of 1812. The survey was undertaken by William Rector, principal deputy surveyor of Missouri Territory, and two field parties: the party surveying north along the meridian was headed by Prospect K. Robbins; that setting the baseline was led by Joseph C. Brown. The surveyors set the baseline from a point at the mouth of the Saint Francis River due west to the Arkansas River, along the Base Line Road in modern-day Little Rock. The principal meridian, that is the Fifth Principal Meridian, began at the mouth of the Arkansas River and ran north to the Missouri River. Along the meridian, Robbins set the section and quarter-section corners of each township from Township 1 North at the intersection of the baseline and the principal meridian to the Missouri boundary.

Given the primitive techniques employed by surveyors of the time, the progress of the initial parties was slow, and the division of the land even slower. While the principal meridian was established by December 28, 1815, the baseline took much longer to complete. In 1818 an additional contract was given to Thomas Rector to extend the baseline toward the western boundary. Not until 1838 did the line reach the eastern boundary of Scott County. The western border was finally reached in 1841.

Despite the time taken to finish the survey, the completion of the work of Robbins and Brown made possible the opening of the lands in much of eastern Arkansas. The first land sales began in 1821.

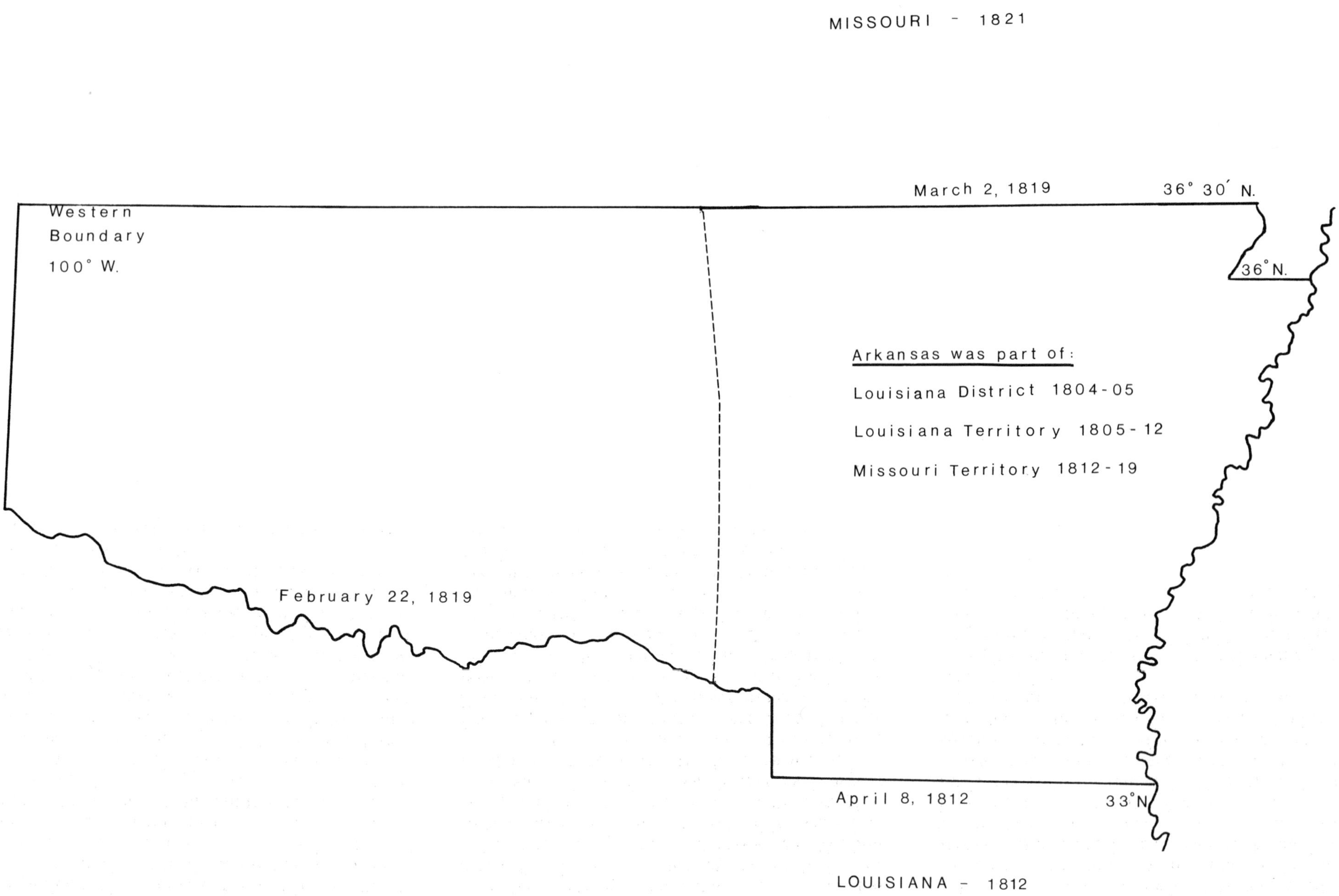

27. TERRITORIAL BOUNDARY CHANGES, 1803–1819

27. TERRITORIAL BOUNDARY CHANGES, 1803–1819

Arkansas became a part of the United States as a part of the Louisiana Purchase in 1803. On March 26, 1804, Congress divided the Louisiana Purchase area into two parts: the Territory of Orleans, lying south of the thirty-third parallel, and the District of Louisiana, or Upper Louisiana, north of it. While Spanish Mexico provided a western boundary for the Territory of Orleans, there was no northern or western boundary defined for the District of Louisiana. Arkansas was part of the District of Louisiana and its capital was at Saint Louis. Initially, however, the government of the district was in the hands of the authorities of the Indiana Territory.

On March 3, 1805, the District of Louisiana became the Territory of Louisiana. President Thomas Jefferson appointed General James Wilkinson as governor of the territory. Wilkinson served until March 3, 1807, when he was replaced by Captain Meriwether Lewis. The territorial legislature and government were still located at Saint Louis. On June 27, 1806, the legislature subdivided the Louisiana Territory, creating a District of Arkansas out of the "southern part of New Madrid District." The boundary was not clearly fixed until August 20, 1808, when Lewis proclaimed the northern boundary as "a line begining on the Mississippi River opposite the second Chickasaw Bluff [at 36° 30′] and continuing west indefinitely." The southern boundary continued to be the thirty-third parallel.

The admission of Orleans Territory as the state of Louisiana on April 8, 1812, established the northern boundary of that state at the thirty-third parallel. On June 4, 1812, Congress changed the name of Louisiana Territory to the Territory of Missouri. In 1812, Governor Benjamin Howard divided the new Missouri Territory into five counties, the old District of Arkansas becoming a part of New Madrid County. On December 31, 1813, however, the territorial legislature created Arkansas County out of what had been the old District of Arkansas, and Arkansas Post was named as its county seat. In 1815 the legislature created Lawrence County, which would become the upper part of Arkansas, out of the southern part of New Madrid County.

In 1818 the territorial legislature created Pulaski, Clark, and Hempstead counties out of Arkansas County. These four counties—Arkansas, Clark, Hempstead, and Pulaski—along with Lawrence would become the basis for the Territory of Arkansas created on March 2, 1819. At this time Arkansas's western boundary was the 100th meridian and the Red River on the south, where the Adams-Onís Treaty of February 22, 1819, had established America's border with the Spanish colonies in the Southwest. Missouri Territory ultimately became a state in 1821.

27. TERRITORIAL BOUNDARY CHANGES, 1803–1819

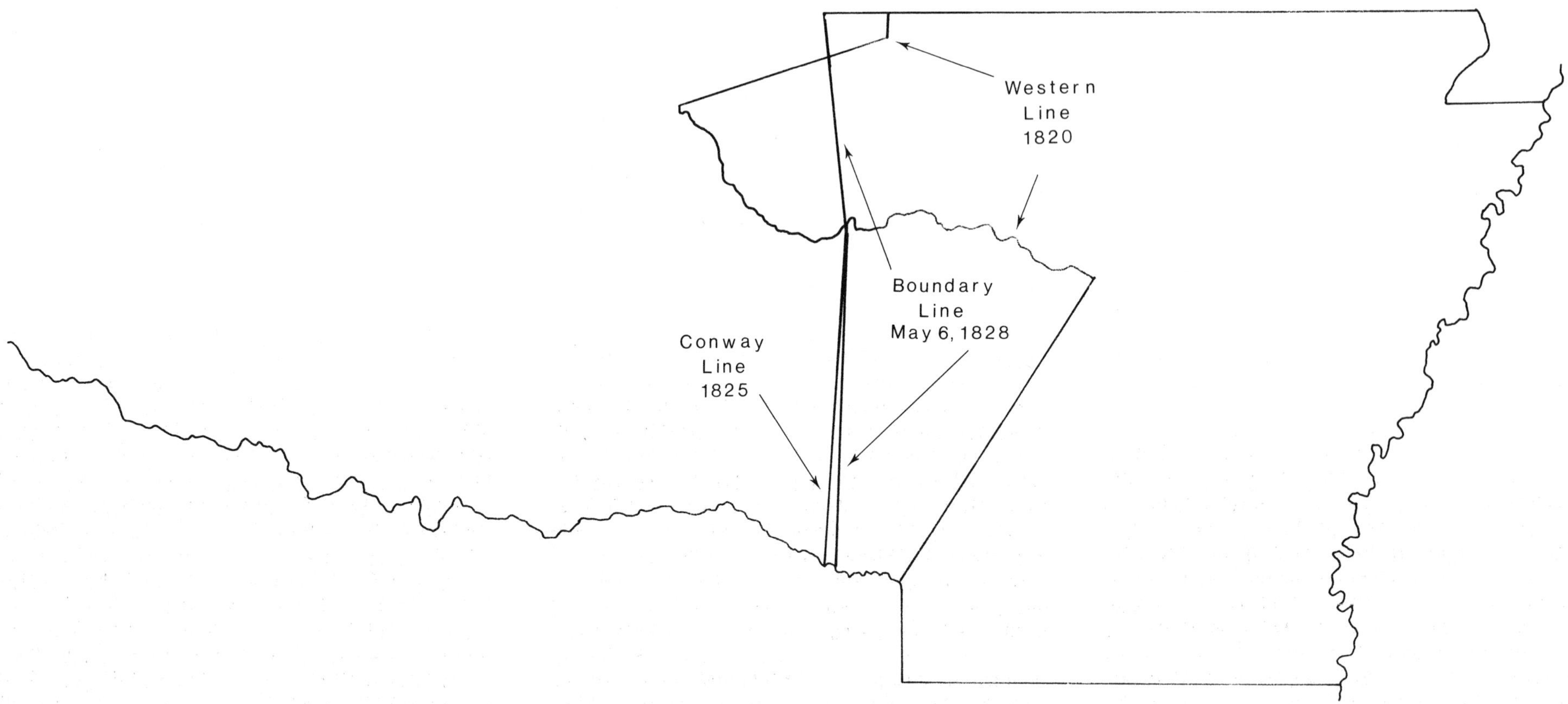

28. TERRITORIAL BOUNDARY CHANGES, 1819–1836

28. TERRITORIAL BOUNDARY CHANGES, 1819–1836

When, on March 2, 1819, President James Monroe signed the bill that created a separate "Arkansaw Territory," the new bill left the southern boundary between Louisiana and the new territory intact and established the present northern boundary of the state of Arkansas, which runs westward from where the thirty-sixth parallel hits the Mississippi River to the Saint Francis River, then north along that river to 36° 30′, then westward to the western territorial boundary at 100° west longitude. The western boundary was not clearly defined in the bill, although the Adams-Onís treaty of 1819, which established the boundary between the Louisiana Purchase territory and Spanish colonies in America, effectively placed the border on the 100th meridian west. The western boundary of Arkansas would be moved several times before statehood.

Continuing problems relating to the western boundary centered on efforts by the federal government to relocate Indians in an Indian Territory west of Arkansas. The Choctaw treaty of 1820 (see Map 20) was received with particular hostility by white Arkansas settlers. Henry D. Downs, who surveyed the western line established by the treaty, found 3,700 white people living in territory assigned to the Choctaw. When the government ordered the removal of these settlers, Congress received numerous protests. The territorial assembly even proposed a new boundary in 1823, to run from the southwest corner of Missouri to the falls of the Verdigris River, to the mouth of the Kiamasha on the Red River.

Protests and pressure finally led Congress to establish a different western boundary in the act of May 26, 1824, that defined the border between Arkansas and the Indian Territory as a line beginning forty miles due west of the southwest corner of Missouri, running due south to Red River, then to the northwest corner of Louisiana. The bill also provided for the negotiation of a new treaty with the Choctaw Indians. This line was known as the Conway Line because of the support given to it by the territorial representative in Congress, Henry W. Conway.

Counter complaints by the Indian tribes effectively prevented the line of 1824 from becoming the permanent western boundary. Two treaties would ultimately bring to an end disputes over the western line. The 1825 Choctaw treaty established a boundary on a north-south line between a point one hundred paces east of Fort Smith and the Red River. As surveyed, the line ran slightly to the west of its intended path. As the result of a later survey the federal government paid the Choctaw for the land taken from them, but the boundary line was not changed. The 1828 Cherokee treaty established the boundary north of the Arkansas River. This line ran from the southwest corner of Missouri to where the Choctaw treaty line intersected the Arkansas River. Like the Choctaw line, the survey of the Cherokee line proved inaccurate, but it remained permanent. The boundaries established in 1825 and 1828 became the state's western boundary in 1836.

28. TERRITORIAL BOUNDARY CHANGES, 1819–1836

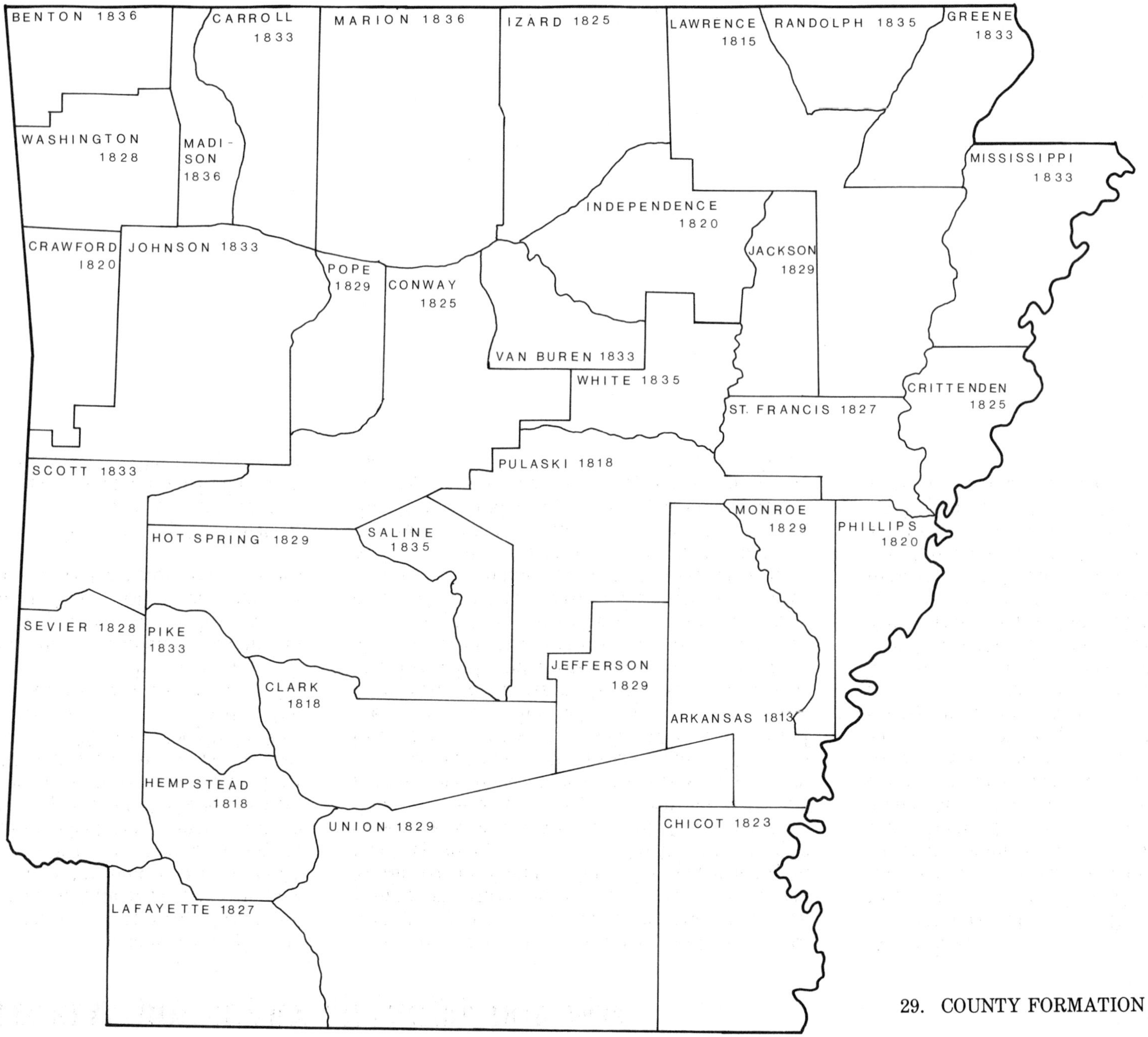

29. COUNTY FORMATION TO 1836

29. COUNTY FORMATION TO 1836

After the establishment of territorial and state boundaries, the organization of counties was important in the creation of the political identity of Arkansas. The creation of the counties reflected the competition for local political power, both in the general population and among specific economic groups. Maps 29, 30 and 31 show the development of such groups in Arkansas. The power to create counties, as well as the governments of cities and towns, belongs to the state legislatures of the United States as a part of their implied authority to create municipal and other corporations. This power has been restricted only by constitutional limits. The first county was laid out in what would become Arkansas on December 31, 1813, when Arkansas County was created south of New Madrid County in the Missouri Territory. By 1836, when Arkansas was admitted to the Union, thirty-five counties would have come into existence.

On January 15, 1815, the General Assembly of the Missouri Territory divided the District of Arkansas into two counties, Arkansas and Lawrence. Lawrence was the area north of the Arkansas River. At the assembly's session of 1818, the district was further divided. On December 15, three new counties, Clark, Hempstead, and Pulaski, were carved out of Arkansas County. After the creation of Arkansas Territory, these five counties remained and served as the basis for representation in the first elected legislature for the territory, which met on February 7, 1820.

In the first special session in 1820, the Arkansas General Assembly created two additional counties—Miller, on April 1, and Phillips, on May 1. When the Assembly convened for another session in the same year, they created the counties of Crawford, on October 18, and Independence, on October 20. The greater portion of the territory of the original Miller County fell within the limits of the state of Texas. It was abolished in 1836, but reestablished in 1874 with the eastern boundary extended. Certain territory was detached from Little River County and added to Miller in 1915.

After 1820 the county boundaries remained fairly stable until 1824. Then between 1824 and 1827 there was a steady expansion of counties. The third General Assembly created Chicot County on October 25, 1823. In 1825 the Fourth General Assembly added Conway on October 20, Crittenden on October 22, and Izard on October 27. In 1827 the Quapaw treaty necessitated a redivision of Arkansas and Pulaski counties. Additionally, the Fifth General Assembly created the counties of Lovely and St. Francis on October 13, and Lafayette on October 15.

On October 17, 1828, in a special session, the General Assembly created Sevier County and abolished Lovely County, although it formed Washington County out of what had been the eastern portion of Lovely County. On November 2, 1829, the Sixth Legislature added Hot Spring, Jefferson, Monroe, Pope, and Union counties to the state. On November 5 they created Jackson County. In 1833 the Eighth Legislature added seven new counties to the state: on November 1, Carroll, Mississippi, and Pike; on November 5, Greene and Scott; on November 6, Johnson; and on November 11, Van Buren.

The final territorial assembly took time in 1835 to create White County on October 23, Randolph on October 29, Saline on November 2, and Searcy (changed in 1836 to Marion) on November 3.

This map includes counties in existence by the end of 1836, and shows Benton and Madison counties, formed by the first General Assembly of the state of Arkansas in 1836.

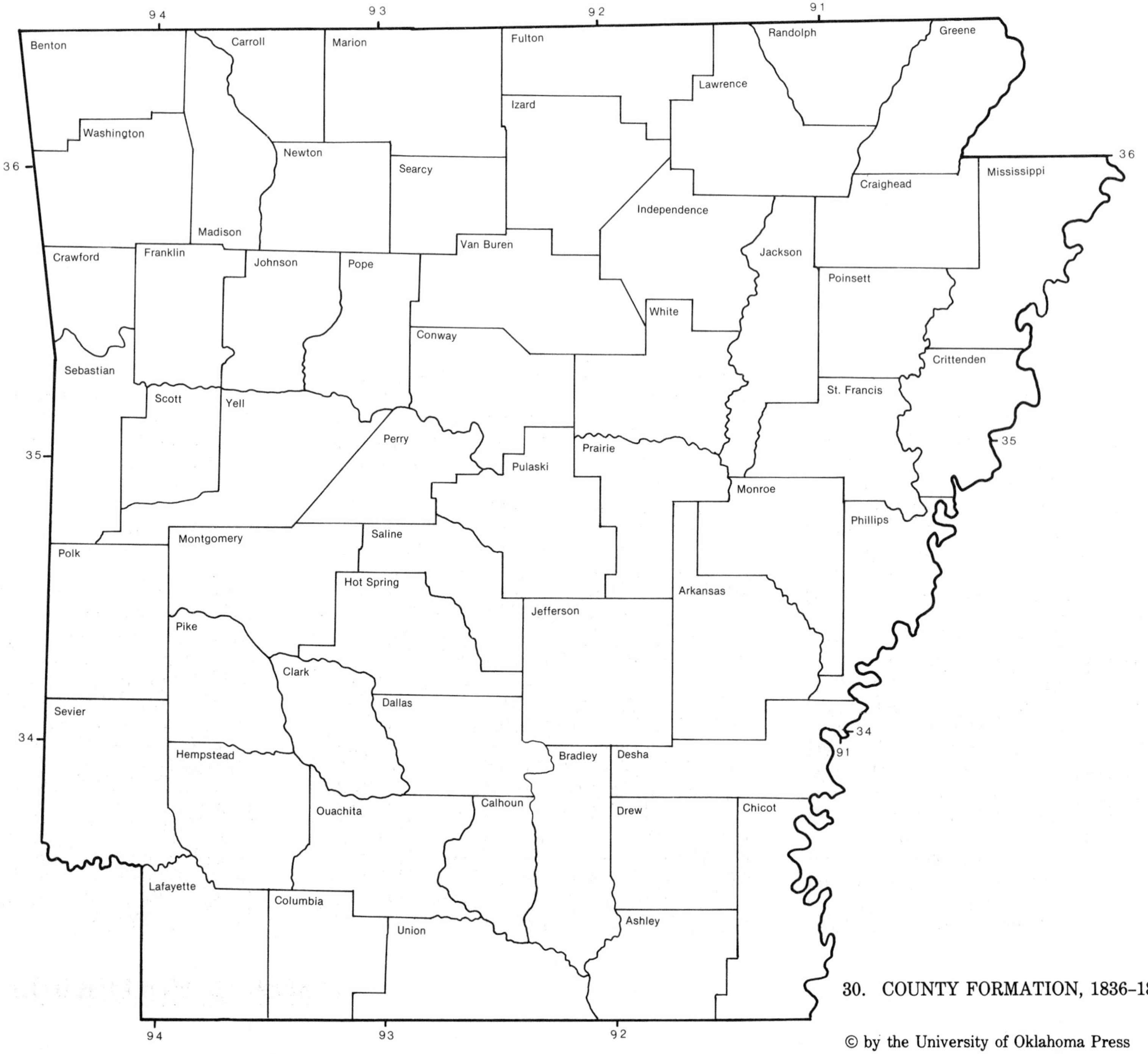

30. COUNTY FORMATION, 1836–1860

30. COUNTY FORMATION, 1836-1860

After Arkansas achieved statehood on June 15, 1836, the creation of counties was subject to the restrictions imposed by article 4, section 29, of the new state constitution. That section read, "No county now established by law, shall ever be reduced by the establishment of any new county or counties, to less than nine-hundred square miles, nor to a less population than its ratio of representation in the House of Representatives; nor shall any county be hereafter established, which shall contain less than nine hundred square miles, (except Washington County, which may be reduced to six hundred square miles,) or a less population than would entitle each county to a member in the House of Representatives." The eleventh amendment to this constitution, ratified in 1859, countered the size requirement by declaring that no counties already established would be unconstitutional should they contain less than nine hundred square miles. The population provision of article 4, section 29, set a requirement for each county to have at least five hundred free white male inhabitants. The proliferation of counties between 1836 and 1860 reflected the continued and rapid growth of the new state.

Under these provisions the legislature set about dividing Arkansas. Between 1836 and 1840 eight new counties were established. Marion County was formed on September 25, Benton on September 30, and Madison on November 30, 1836. Franklin was created on December 19, 1837. In 1838 two counties were established: Poinsett, on February 28, and Desha, on December 12. Yell County was created on October 5, Bradley on October 18, and Perry on December 18, 1840.

Between 1840 and 1850 ten counties were created. In 1842, Ouachita County was formed on November 29, Montgomery County on December 9, Newton on December 14, and Fulton County on December 21. Polk was created on November 30, 1844; Dallas County, January 1, 1845; Prairie, November 25, and Drew, November 26, 1846; Ashley, November 30, 1848; and Calhoun, December 6, 1850.

From 1850 to 1860 three new counties were formed: Sebastian, on January 6, 1851; Columbia, on December 17, 1852; and Craighead, on February 19, 1859.

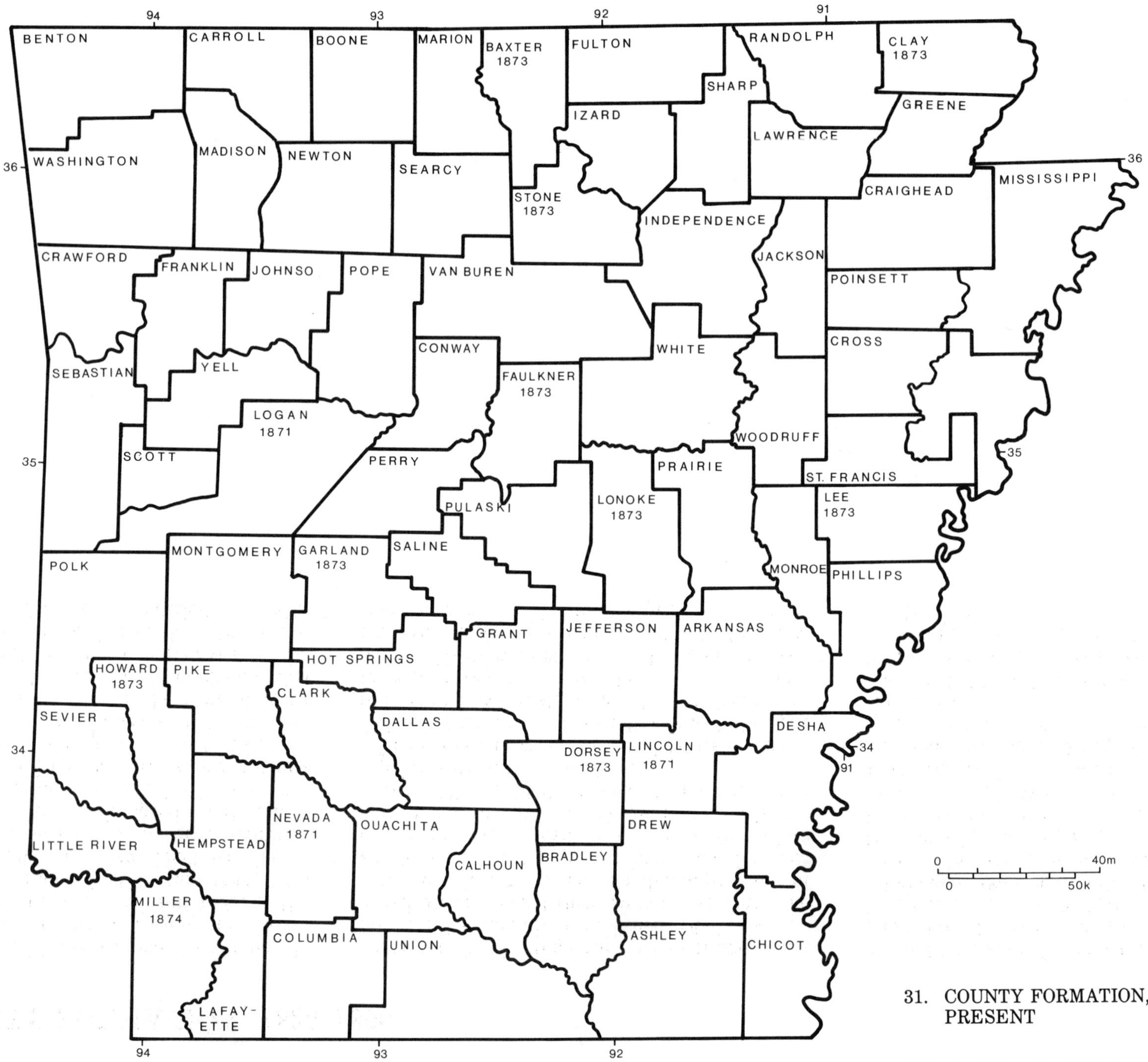

31. COUNTY FORMATION, 1861 TO THE PRESENT

31. COUNTY FORMATION, 1861 TO THE PRESENT

Between 1860 and the creation of the last county in Arkansas in 1883, four constitutions provided the framework for the organization of new counties. The first of these was the constitution of the government organized after secession in 1861. Only two counties were created by the Confederate legislature. Cross County, named after Poinsett County land speculator David C. Cross, was organized on November 15, 1862. Woodruff County, named after William E. Woodruff, founder of the *Arkansas Gazette,* was established on November 26, 1862. The legality of the actions creating each of these counties had to be approved after the Civil War.

In 1864, Arkansas Unionists wrote a constitution that lowered size restrictions on new counties while keeping the population requirement at a minimum of five hundred adult white male inhabitants. Section 27 of the new constitution said, "No county now established by law shall ever be reduced, by the establishment of any new county or counties, to less than six hundred miles nor to less population than its ratio of representation in the house of representatives, nor shall any county be hereafter established which shall contain less than six hundred square miles or a less population than would entitle each county to a member in the house of representatives." Little River was created on March 5, 1867, and Sharp on July 18, 1868—the only two counties organized under the 1864 constitution.

In 1868 a new constitution was written that placed no restrictions on the size or population of new counties. Boone County was created on April 9, and Grant County on February 4, 1869; Nevada on March 20, Sarber on March 22, and Lincoln on March 28, 1871; Baxter and Clayton Counties on March 24, Garland on April 5, Faulkner county on April 12, Lonoke on April 16, Lee, Howard, and Dorsey counties on April 17, and Stone on April 21, 1873. Clayton, Dorsey and Sarber counties were named after prominent Arkansas Republican politicians. After Reconstruction the Democratic General Assembly of 1875 changed Clayton County to Clay and Sarber to Logan, and in 1885, Dorsey County was changed to Cleveland in honor of the first Democratic President elected after the Civil War.

In 1874 a new constitution was adopted. In article 13, section 1, the document stated, "No County now established shall be reduced to an area of less than 600 square miles nor less than five thousand inhabitants, nor shall any new county be established with less than six hundred square miles and five thousand inhabitants. Provided, that this section shall not apply to the Counties of Lafayette, Pope and Johnson, nor be so construed as to prevent the general assembly from changing the line between the Counties of Pope and Johnson." Section 2 went on to say, "No part of a county shall be taken off to form a new county or a part thereof without the consent of a majority of the voters in such part proposed to be taken off." Under the new constitution only one final county was created in the state, and that was Cleburne County, named after Confederate general Patrick R. Cleburne, created on February 20, 1883.

31. COUNTY FORMATION, 1861 TO THE PRESENT

PART FIVE
EARLY ARKANSAS

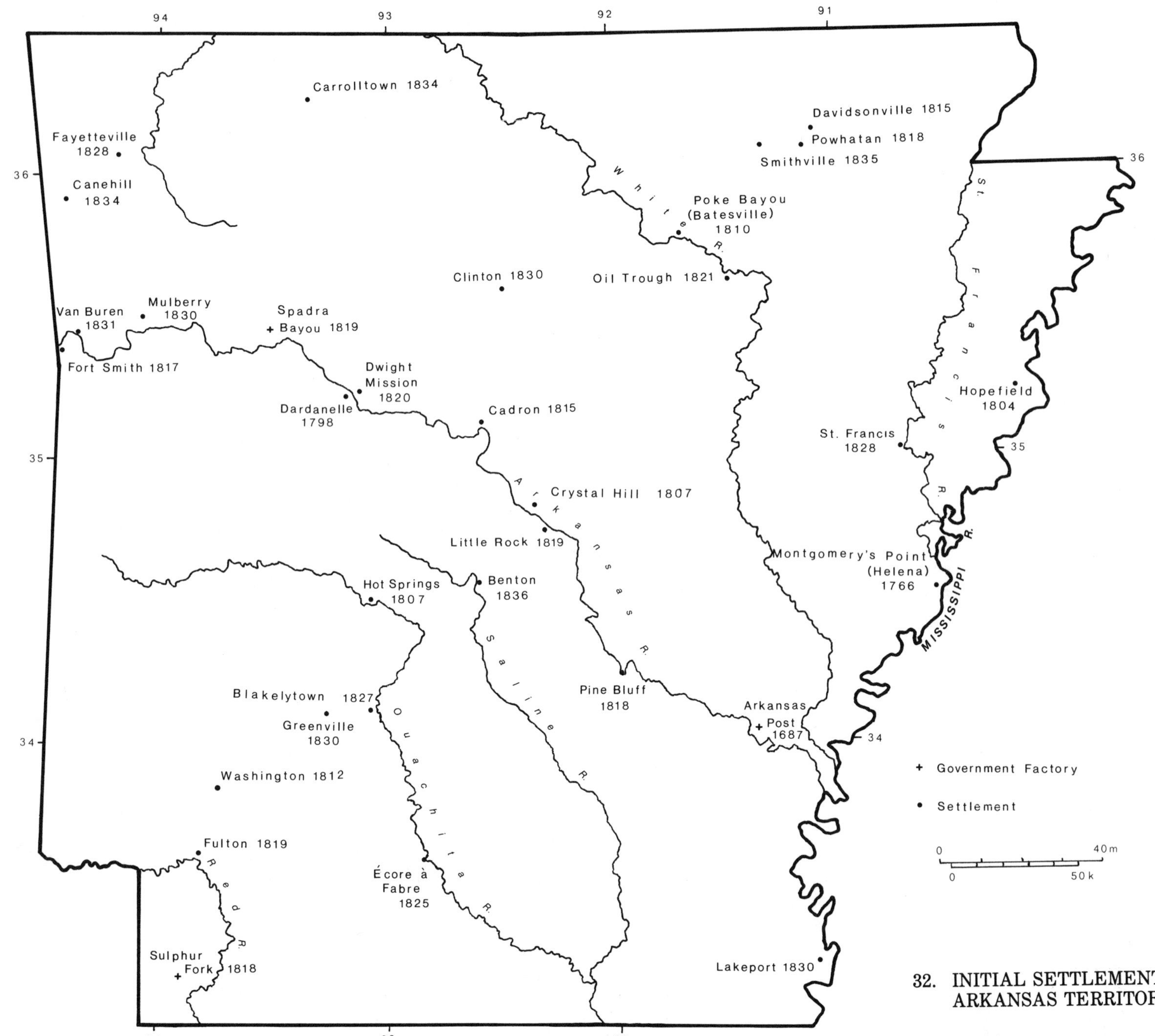

32. INITIAL SETTLEMENT PATTERNS IN ARKANSAS TERRITORY

32. INITIAL SETTLEMENT PATTERNS IN ARKANSAS TERRITORY

The pattern of settlement among the first white inhabitants of Arkansas reflected the routes taken by them as they entered the state, traveling by canoe, on horseback, or on foot. Geographically, the area contained many barriers to transportation, beginning with the swamps of the eastern rivers and ending with the rugged terrain of the western mountains. These barriers funneled immigration and early trade into narrow lines around them and thus helped determine the earliest development of the area.

The rivers were the obvious pathways through and into Arkansas, and consequently the earliest settlements were located near them. Communities developed slower away from the rivers. The Mississippi River provided the first road for whites to Arkansas, but settlement along the Mississippi was delayed because much of the land on the Arkansas shore was not suitable for town growth. Except for several miles of highlands at the southern end of Crowley's Ridge, at present-day Helena, the Mississippi shores on the Arkansas side were subject to frequent flooding. As a result, the first permanent white communities were along the Arkansas River, upriver from its mouth on the Mississippi. In 1686, Henri de Tonti established Arkansas Post on the lower Arkansas River, and, although abandoned at various times, it remained a point of settlement into the nineteenth century.

Further development along the Arkansas was slow in taking place, although it is clear that American settlers began to move into the river valley soon after the Louisiana Purchase. In 1810 a special census showed 765 white, 2 free colored, and 107 slave inhabitants along the Arkansas. Several small communities developed in the vicinity of modern Little Rock. One of the earliest was Crystal Hill on the north bank of the river west of Little Rock, settled by John and Jacob Pyeatt in 1807. It was also called Pyeattsville, although the name Crystal Hill persisted. In 1815 a group headed by Alexander McFarland settled what would become Cadron, at the mouth of Cadron Creek. Little Rock, on the south bank of the river, had been the site of earlier settlements that began in 1819 and led to the incorporation of Little Rock in 1831. Pine Bluff was settled as early as 1818, but a town was not laid out there until 1832. Upriver the federal government in 1817 created a post at Fort Smith which was later the basis of a community, and by 1831, Van Buren had developed across the river. In the area between Little Rock and Fort Smith on the Arkansas settlers had arrived by 1798 at Dardanelle, which was later the site of the Cherokee Indian Agency; Cephas Washburn established Dwight Mission for the Cherokee on Illinois Bayou, in modern-day Pope County, in 1820; and settlers were at Mulberry by 1830.

Settlement along the Mississippi did not take place until eighty years after the creation of Arkansas Post. The second permanent settlement in Arkansas emerged in 1766 on the Mississippi River at Montgomery's Point, below the mouth of the Saint Francis River, where Francis d'Armand had established a French trading post. A town, Saint Francis, developed near that location and was later the site of Helena. In 1797 the Spanish established a post up the Mississippi that they called Fort Esperanza, which was located across from modern-day Memphis, Tennessee. American settlement of that site began when Benjamin Foy moved there in the same year. He established what would become the village of Hopefield in 1804. Enumerators who gathered the Missouri Territory's census in 1810 counted a combined population of 159 whites and 29 slaves in Hopefield and Saint Francis. On the south, a settlement had been made at Lakeport, in present-day Chicot County, by 1830.

Some settlers also moved into Arkansas on an Indian trail that lay along the edge of the Ozark and Ouachita highlands, running from the northeast to the southwest. The terrain offered a natural roadway, dry and relatively level. The route would be developed later and known as the Southwest Trail. A line of small villages and individual farmsteads developed on or near it. The first settlement encountered upon entering Arkansas Territory was established by Kentuckians in 1810 on the White River in present-day Independence County, near the site of Batesville. Other settlements in that vicinity included Powhatan (1818), Davidsonville (1815), Oil Trough (1821), and Smithville (1835). The trail passed through the settlements near present-day Little Rock. On the southwest, communities emerged at Hot Springs (1807); Blakelytown or Hemphill and Greenville (1830), near modern-day Arkadelphia; Washington, settled by John Campbell of Tennessee (1812); and Fulton, on the Red River (1819). South of this path, Camden developed at Écore à Fabre on the Ouachita River.

The northwest prairies and Ozark region were slower to settle. By 1828, however, a community was at Fayetteville. Canehill and Carroltown followed in 1834. Benton had begun to develop by 1836. A town had emerged at Clinton by 1830.

This spread of settlements reflected the steady growth of Arkansas under the control of the United States. The special census in 1810 reported a population of only 1,062 inhabitants. By 1820 the number had increased to 14,255. In 1830, on the verge of statehood, the region contained a population of 30,388.

32. INITIAL SETTLEMENT PATTERNS IN ARKANSAS TERRITORY

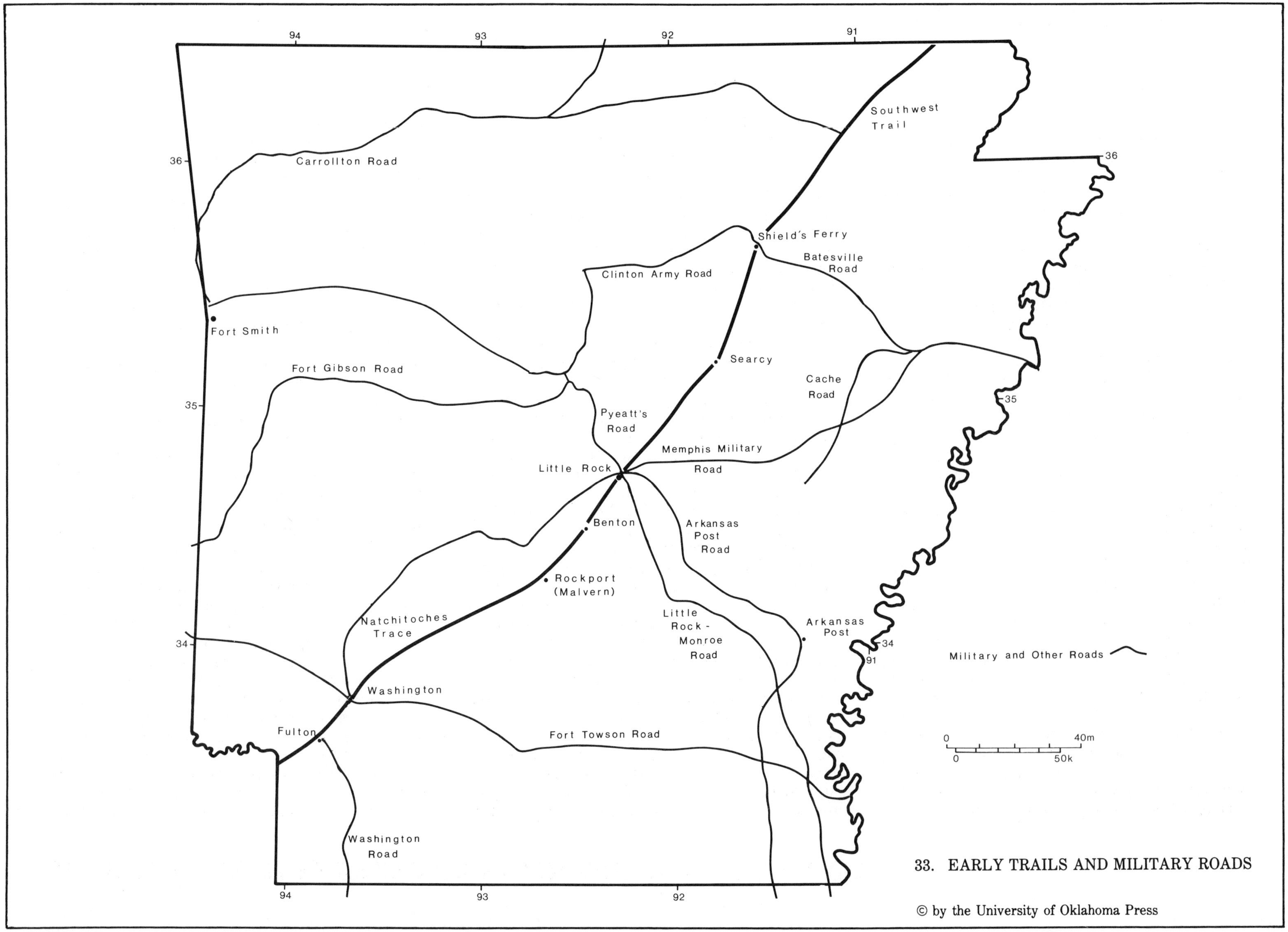

33. EARLY TRAILS AND MILITARY ROADS

33. EARLY TRAILS AND MILITARY ROADS

As more settlers entered Arkansas, the system of roads began to expand as the immigrants attempted to overcome the basic geographic barriers to their movement and communication. As they moved away from the rivers, the new Arkansans began to develop a system of roads, but at first even these had to follow the routes that nature made possible. Many of the early roads followed Indian routes through the area. At the beginning these were simple wilderness trails, often marked only by blazed trees, with no improvements. S.A. Mitchell's map of 1836 is the best source of information on land routes at the time of statehood and shows the earliest developed overland routes. There were many roads in the state, but those described here were the most heavily traveled and the most important.

Immediately after the Louisiana Purchase in 1803, Congress appropriated funds to improve the Indian trail that ran across Arkansas from southeastern Missouri to Fulton on the Red River, and to incorporate it into the nation's Southwest Road, which began in Illinois and ran to Texas. The money was to be used to build a road that could be used by wagons, and it was to have a branch to Monroe, Louisiana. Development, however, was slow. As laid out, the road known as the Southwest Trail began in Illinois and ran south, entered Arkansas in what is now Randolph County, crossed the White River at either Shield's Ferry above Batesville or Russell's Ferry below, then proceeded southwest to near modern-day Searcy. From that point it went on a direct line to Little Rock. From Little Rock the road ran to Benton, Rockport on the Ouachita, Washington, and Fulton on the Red River.

Another of the early roads, known as Pyeatt's Road, was constructed in 1807 under the direction of Major John Pyeatt of Crystal Hill, near Little Rock. The road ran directly from central Arkansas along the north bank of the Arkansas River for about fifty miles below Crystal Hill, then struck off in a southeasterly direction along an old Indian trail to Arkansas Post. From the Post the road ran on to the Mississippi.

These major arteries were supplemented by many others. As early as 1816 the Natchitoches Trace ran from Natchitoches in Louisiana to the Southwest Road, to Hot Springs, then on to the vicinity of Little Rock on the Arkansas River. Other major routes included a trail that ran from Point Chicot to Camden to Washington, where it connected with the Southwest Road. This road later would be called the Fort Towson Road. In the north, a path known as the Carrollton Road existed from the town of Jackson in Lawrence County and ran westward to Fayetteville and then to Fort Smith. Another road, later known as the Clinton Army Road, ran from Batesville to Clinton to Morrilton. The Little Rock–Monroe Road went south from Little Rock to Pine Bluff, then to near McGehee, then to Monroe, Louisiana. A final major route was the Batesville Road, which went from the Saint Francis River to near Clarendon, then to Batesville.

Improvements on these trails began in 1824, when the first military road was constructed in the state. Congress appropriated fifteen thousand dollars for the Memphis Military Road, backed by Henry W. Conway, from Memphis to Little Rock. Development was to be extended later to Fort Smith. The survey of the proposed route was completed in 1826, and the first contracts were let in the same year. The contracts provided for a path twenty-four feet wide with timber and other impediments removed, though stumps were not removed. This road was completed in 1828, but it remained in bad shape throughout the antebellum years. Catholic Bishop John B. Morris of Little Rock, who traveled along the route in 1836, reported that much of the roadway was either under water or so muddy that his coach averaged only three miles an hour. The western portion of the road was completed in 1827.

Under Andrew Jackson the Southwest Trail was designated a military road from where it entered Arkansas in the northeast to Fulton on the Red River. After 1831 this trail became part of the National Road, which began in Cumberland, Maryland, and had reached Illinois. However, the change in status did little for the condition of the Trail, as was also true of the road from Memphis to Fort Smith. When George W. Featherstonhaugh traveled along the Southwest Trail in 1834, he noted that the way was full of rocks and stumps. Fallen trees that had not been removed, missing bridges, and mud impeded his progress.

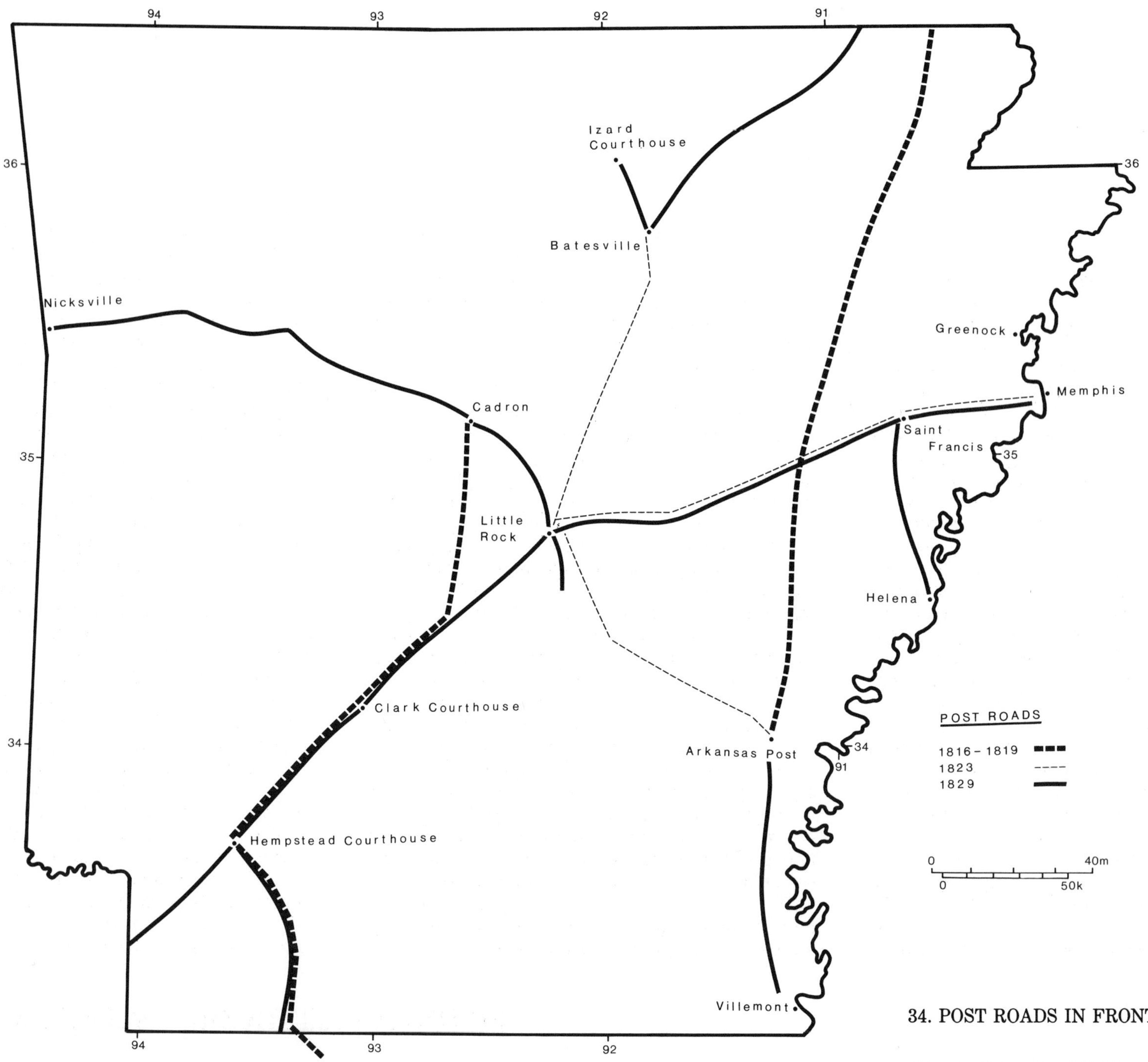

34. POST ROADS IN FRONTIER ARKANSAS

34. POST ROADS IN FRONTIER ARKANSAS

After 1816 the establishment of postal routes in Arkansas led to a further development of overland transportation and improved communication. Unfortunately, the government put no money into the construction of such roads, and they were usually little more than trails through the forests and swamps of Arkansas. These roads did, however, tie Arkansans together more closely and increased contact among the area's inhabitants.

On April 30, 1816, Congress established the first post road in the Arkansas Territory from Saint Louis to Arkansas Post. On March 3, 1819, another post road was established between Cadron and Monroe, Louisiana, on a route that went through Clark Courthouse and Hempstead Courthouse along the Natchitoches Trace. Mail service was restricted to one delivery per month.

In 1823 the government let contracts on three new routes. These included weekly mail from Batesville to Little Rock, from Arkansas Post to Little Rock, and from Memphis to Little Rock. In 1826 the road between Little Rock and Arkansas Post was in good enough shape that Wright Daniels put the first stagecoaches in the territory in operation to carry passengers and the mail. The 120-mile trip took over forty-eight hours. Although the postmaster general intended to use coaches on the Memphis–Little Rock route as early as 1827, this service did not begin until 1831.

By 1829 increases in population had forced an expansion of the postal system. At the time the service had grown to include at least fourteen routes. Among the services provided was a weekly mail between Memphis and Little Rock over the military road, a weekly mail between Memphis and Greenock in the northeastern corner of Crittenden County, mail every two weeks between Batesville and Izard Courthouse, weekly mail from Little Rock to Nicksville near modern-day Van Buren, weekly mail between Saint Francis and Helena, weekly mail between Little Rock and Miller Courthouse in Texas, mail every two weeks between Hempstead and Natchitoches, mail every two weeks between Arkansas Post and Villemont on the Mississippi River in Chicot County, weekly mail between Little Rock and Monroe, Louisiana, and weekly mail between Greenville, Missouri, and Batesville. Most of the routes into the west were poorly developed, and it was not until 1838 that the Beall and Richmond Company established a mail stage between Little Rock and Washington, in Hempstead County. By 1840 a similar line traveled between Fort Smith and Little Rock.

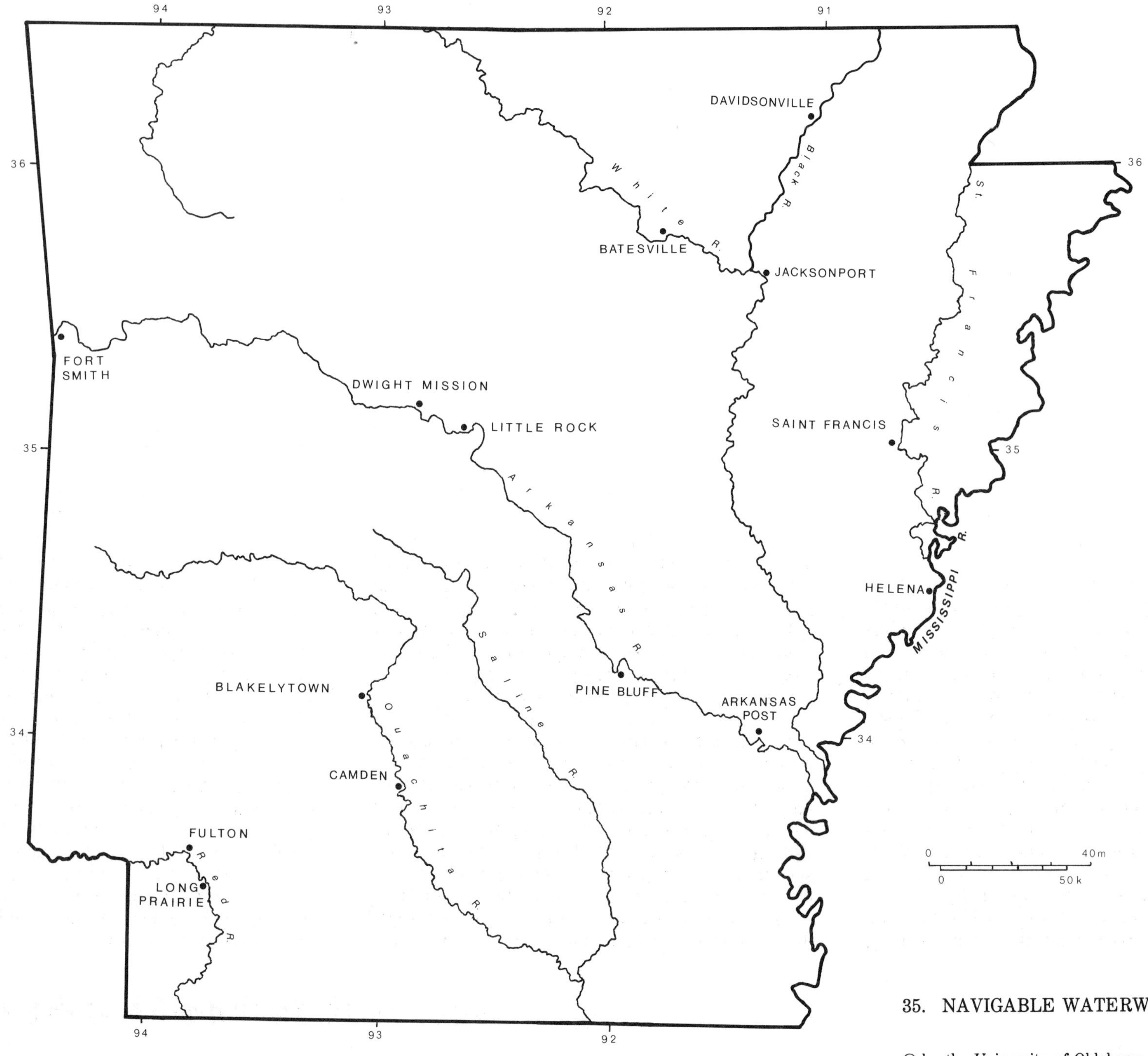

35. NAVIGABLE WATERWAYS, 1860

35. NAVIGABLE WATERWAYS, 1860

By 1820 steamboats appeared on Arkansas rivers, providing a more reliable, safer, and cheaper means of transportation than had been available before. The failure to improve the road system through the antebellum years meant that the riverways became the chief avenues of travel and commerce. The highly versatile river steamboat could go almost anywhere and would continue to be the chief carrier of the state's products and peoples until it was supplanted later in the nineteenth century by the railroad. Commerce and communication thus remained confined to natural routes. Many of Arkansas Territory's early towns already lay along river courses when the new invention arrived in Arkansas. Many of them now prospered, as the steamboats turned towns such as Little Rock, Pine Bluff, Camden, and Helena into major ports and into the region's commercial centers.

The first steamboats operated along the Arkansas River, where, on March 31, 1820, the *Comet* became the first to enter the river's channel and landed at Arkansas Post. On March 16, 1822, the *Eagle* steamed upriver to become the first boat to reach Little Rock, then carried supplies farther west to within ten miles of Dwight Mission. Generally Little Rock was the highest point on the river that could be reached regularly, but with high water the Arkansas could be navigated as far west as Fort Smith in Arkansas and on to Fort Gibson near present-day Muskogee, Oklahoma. On April 25, 1822 the *Robert Thompson* returned to Little Rock from a round trip to Fort Gibson that had taken fifteen days. The steamboat trip above Little Rock, however, was a risky one, and low water could ground a boat upriver for months. Typical problems with water in the river were seen through the middle of the 1850's, when inadequate rainfall on the headwaters of the Arkansas isolated the western portion of the state for almost three years.

The principal avenues to the southern part of the state were along the Ouachita and up the Red River. The Ouachita generally was navigable to Camden. The first steamboat on the Ouachita was the *Dime,* which travelled between Blakelytown, near modern-day Arkadelphia, and New Orleans in the 1830's. At first the Red River was open only as far as the Great Raft, near present-day Shreveport, Louisiana, but as early as 1831 the first steamboat passed through this natural blockade when the *Enterprise* ran up the Red as far as Long Prairie in Lafayette County, Arkansas. Removal of the raft began in 1835, and by March 13, 1838, the first regular commercial traffic reached Fulton with the arrival of the *Indian.*

The White River was the fourth major riverway. Generally it could be navigated as far as Jacksonport, although at times it was open to Batesville. In 1831, Captain Philip Pennywit's *Waverly* became the first steamboat to reach Batesville.

Some commercial traffic also moved on the Saint Francis and Black rivers, although these eastern rivers were slower to be developed than some of the others. In 1831 the *Bob Handy* became the first steamboat to navigate the Black as far north as Davidsonville in Lawrence County. In 1836 the *Gladiator* steamed up the Saint Francis to the town of Saint Francis.

The steamboats remained an important component of the state's transportation network through the nineteenth century, although they lost increasing amounts of traffic to the railroads as the latter developed an operating network by the end of the century.

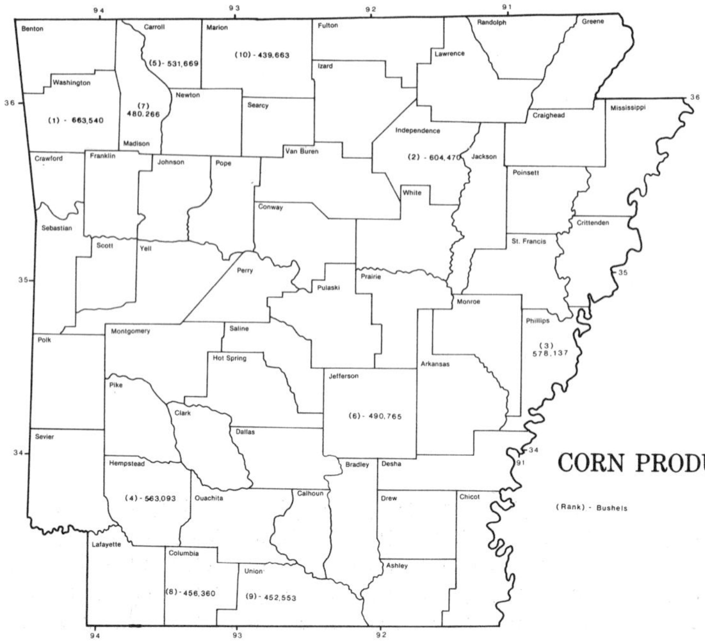

CORN PRODUCTION, 1860

(Rank) - Bushels

COTTON PRODUCTION, 1860

(Rank) - Bales

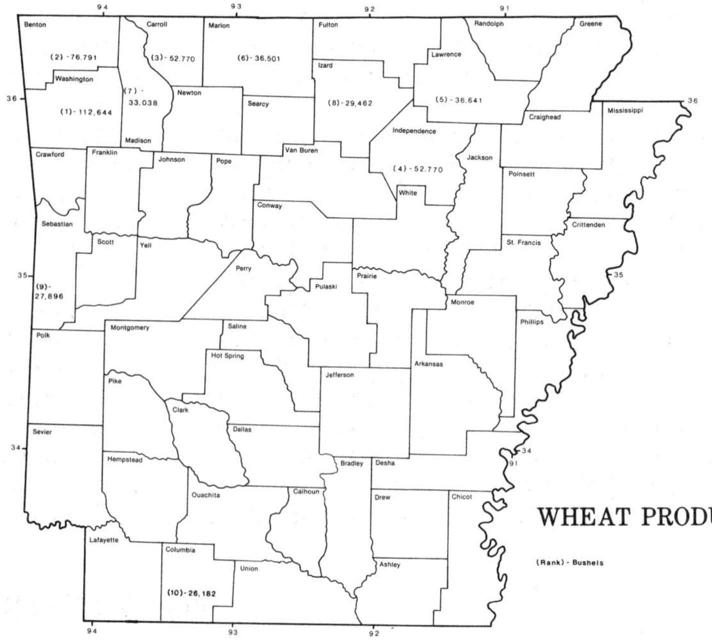

WHEAT PRODUCTION, 1860

(Rank) - Bushels

36. ANTEBELLUM AGRICULTURE

36. ANTEBELLUM AGRICULTURE

The early economy of the Arkansas area centered around such enterprises as trapping and hunting. Agriculture, however, quickly emerged as the principal endeavor of the majority of the residents of the Arkansas Territory and then the state. By 1850 the state's farms produced a variety of products, including livestock for sale and slaughter, wheat, rye, corn, oats, rice, tobacco, cotton, vegetables, and fruit, as well as other crops. The three most valuable crops were cotton, corn, and wheat. Although they did not consider corn a cash crop, farmers everywhere in the state grew it to provide food for themselves and for their livestock; corn was essential for self-sufficiency. A ready national market existed for cotton, and prices for it were high, especially in the 1850's. Wheat, milled into flour, generally moved in local markets and could not demand as good a price as cotton.

Geography helped to create sectional economies in antebellum Arkansas. Cotton was the cash crop of preference, but most of its production took place in the southern and eastern parts of the state, where production was facilitated by good soils and access to markets via navigable rivers. The soils were adequate in the valleys and on the plateaus in northwest Arkansas, and the growing season was long enough, but overland transportation problems discouraged the growth of cotton there. As a result, northwestern Arkansas farmers looked to wheat as their cash crop. The differences in crops created important differences in the wealth and the societies of the two regions.

By 1860 the basic regional agricultural pattern had been set. Corn was grown throughout the state in substantial quantities, as indicated by the counties that produced the largest amounts. Beginning with the largest producer, the top ten counties included Washington, Independence, Phillips, Hempstead, Carroll, Jefferson, Madison, Columbia, Union, and Marion. While all counties in the state except Benton grew some cotton, most of the production was focused in the south and east. The ten principal cotton producers in 1860, in order, were Chicot, Jefferson, Phillips, Marion, Arkansas, Lafayette, Union, Hempstead, Columbia, and Desha. The top five of these produced 37.5 percent of the total crop. Wheat flourished in the northwest. The top ten counties producing wheat were Washington, Benton, Carroll, Independence, Lawrence, Marion, Madison, Izard, Sebastian, and Columbia.

Production and Value of Principal Crops, 1850 and 1860

	1850		1860	
Crop	Amount	Value	Amount	Value
Corn	8,893,939 bu.	$5,692,121	17,823,588 bu.	$11,585,332
Wheat	199,639 bu.	152,101	957,601 bu.	1,431,613
Cotton	65,344 bales	3,274,924	367,393 bales	16,165,292

Note: Figures for corn are based on average New Orleans prices for 1850 and 1860. Other prices are from U.S. Bureau of the Census, *Historical Statistics of the United States, Colonial Times to 1970, Bicentennial Edition* (Washington, D.C. Government Printing Office,1975).

37. MIGRATION INTO ARKANSAS, 1830–1860

37. MIGRATION INTO ARKANSAS, 1830–1860

As the agricultural frontier opened up in Arkansas, the state became an important destination for immigrants. They helped produce a steady population growth during the years before the Civil War. New lands always attracted people from the older states, and Arkansas offered that commodity, plentiful and cheap, at the same time as a combination of economic panics and good prices for cotton encouraged eastern emigrants searching for new opportunities. In the 1830's the state's population growth rate was 221 percent, then 115 percent in the 1840's, and 107 percent in the 1850's. Between 60 and 75 percent of the increase during each of these decades was the result of immigration rather than natural growth.

During these antebellum decades over 90 percent of the new settlers came into Arkansas from other slaveholding states. As in most other regions, the white settlers moved west into Arkansas along horizontal lines into geographical areas similar to those that they left. Before 1850, Tennessee, Alabama, Missouri, and Mississippi accounted for 74 percent of the white families that moved into the state. Tennessee was dominant with 36 percent of the immigrants. This basic trend continued in the 1850's when Tennessee, Mississippi, Alabama, and Missouri were the four principal points of origin for the Arkansas immigrants. In the 1850's one section of the state, the southern and southwestern counties of Clark, Columbia, Union, Montgomery, Pike, and Polk, attracted a large proportion of Georgians.

These people usually sought homes much like those they were leaving. Upland inhabitants of the older states settled for the most part in the Arkansas uplands. Lowland dwellers moved into the Arkansas lowlands. Frequently, the immigrants moved from an adjacent state into the closest county in Arkansas, then moved into the interior, county by county, until they found the land, the climate, and the community in which they believed that they could prosper and be comfortable.

In terms of sources, the black migration before the Civil War generally paralleled that of the white population, since the blacks came as the slaves of white settlers. The majority of the slaves, however, were members of holdings of twenty or more people who represented a major investment by their owners. Such groups were brought into the delta counties of the east and south for use on commercial plantations. Another difference between white and black immigrants was the appearance of a disproportionate number of slaves from the older eastern states, suggesting the existence of a considerable slave trade between the frontier and the eastern regions. For example, in the 1850's over 22 percent of black immigrants came from Tennessee, over 18 percent from Mississippi, and 18 percent from Alabama—as would be expected, considering white trends—but almost another 10 percent came from South Carolina.

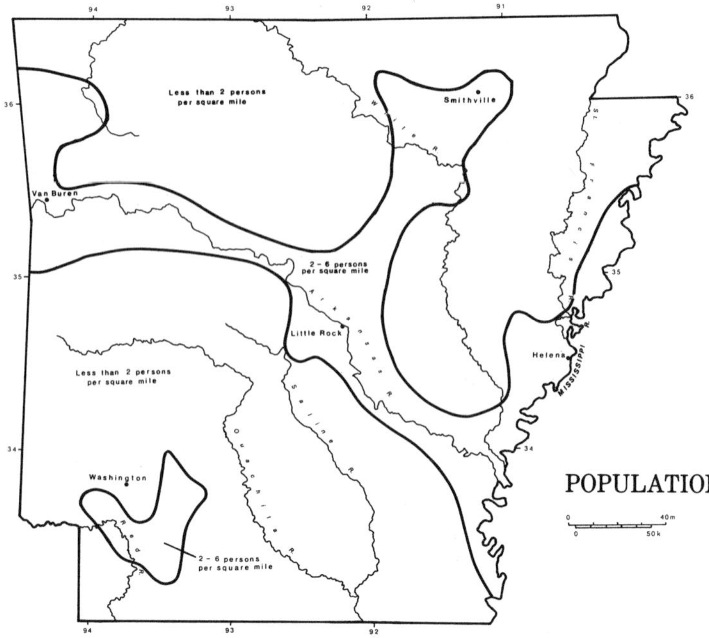

POPULATION DENSITY, 1830

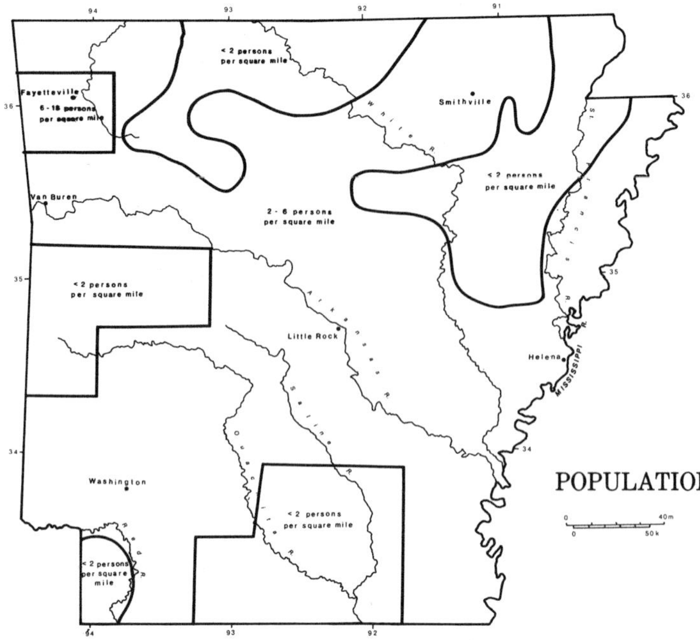

POPULATION DENSITY, 1840

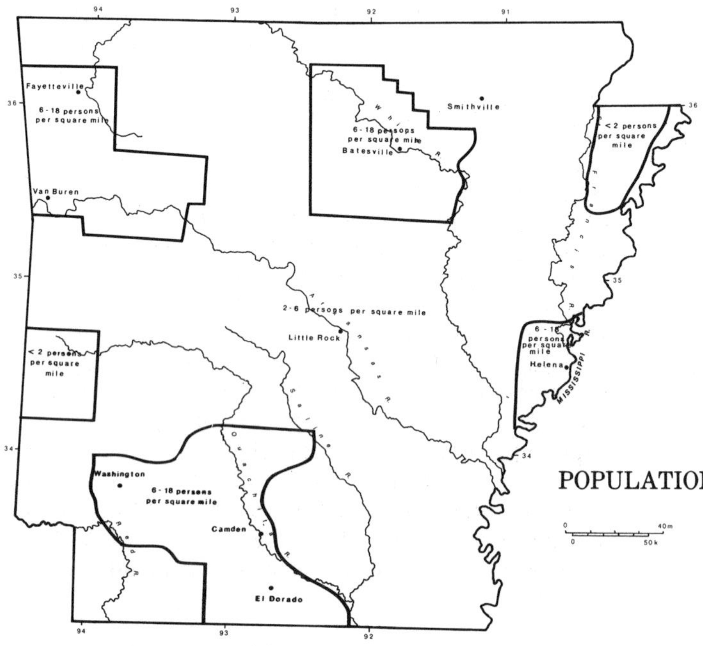

POPULATION DENSITY, 1850

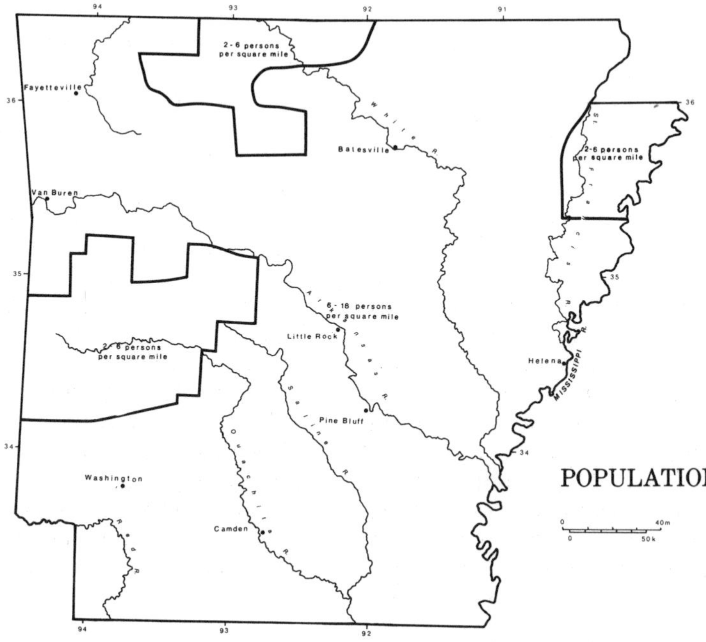

POPULATION DENSITY, 1860

38. SETTLEMENT IN FRONTIER ARKANSAS, 1830–1860

38. SETTLEMENT IN FRONTIER ARKANSAS, 1830–1860

Expansion of transportation and communication and economic development spurred migration and population growth. Settlement patterns reflected the importance of transportation routes and economic opportunities to the people of the area. By 1830 the population of Arkansas had reached 30,388, of which 4,576, or 15 percent, were slaves. Population growth at this time basically reflected the development of Euro-American society in communities along the Arkansas River and the Southwest Trail out of Missouri. Three of the denser population centers were along the trail. Smithville and Lawrence County was one of the first stops. Little Rock and Pulaski County sat astride both the trail and the Arkansas River. Washington in Hempstead County was on the trail and also in the midst of rich, rolling prairie farmlands. A fourth population center was Van Buren in Crawford County, where good agricultural lands and opportunities for trade with the Indian Territory attracted settlers.

A special census taken in 1835 reported that the population of Arkansas Territory had reached 52,240, of which 42,302 were white and 9,938 were black (Indians were not counted in the federal or state censuses). This census showed the population to be more than that required by the federal government for statehood.

By 1840 the new state's population had increased to 97,574. Approximately 21 percent of these people were slaves. The same general settlement trends had continued, although the 1830's brought a surge of population in the northwest, where the settling of land disputes with the Cherokee Treaty of 1828 and good lands encouraged immigration. Fayetteville, in Washington County, was the center of this growth. Centers of population in the 1830's continued to grow and to profit from their earlier advantages.

By 1840 the development of plantation agriculture added a new dimension to the pattern of settlement. Of 209,847 inhabitants, nearly 23 percent were slaves. After 1850 counties in the southern parts of the state showed the greatest growth, as planters began to move in and open plantations along the Ouachita River and in the east along the Mississippi. El Dorado, in Union County, was the focus of southern development. By 1850 that county was the most populous in the state with 10,298 inhabitants, half of whom were slaves. Other areas continued to grow, however, and by 1850 the region around Batesville, in Independence County, had become one of the most heavily populated areas of the state.

By 1860, Arkansas had 435,450 inhabitants, and the proportion of slaves continued to grow. The 111,259 slaves in 1860 represented over 25 percent of the population. Reflecting the rapid expansion of the plantation during this period was the development of denser populations in the vicinity of Pine Bluff, in Jefferson County, and Helena, in Phillips County.

Population Growth, 1810–1860

Year	White	Black	Total
1810	924	138	1,062
1820	12,597	1,676	14,273
1830	25,671	4,717	30,388
1840	77,174	20,400	97,574
1850	162,189	47,708	209,897
1860	324,143	111,259	435,450

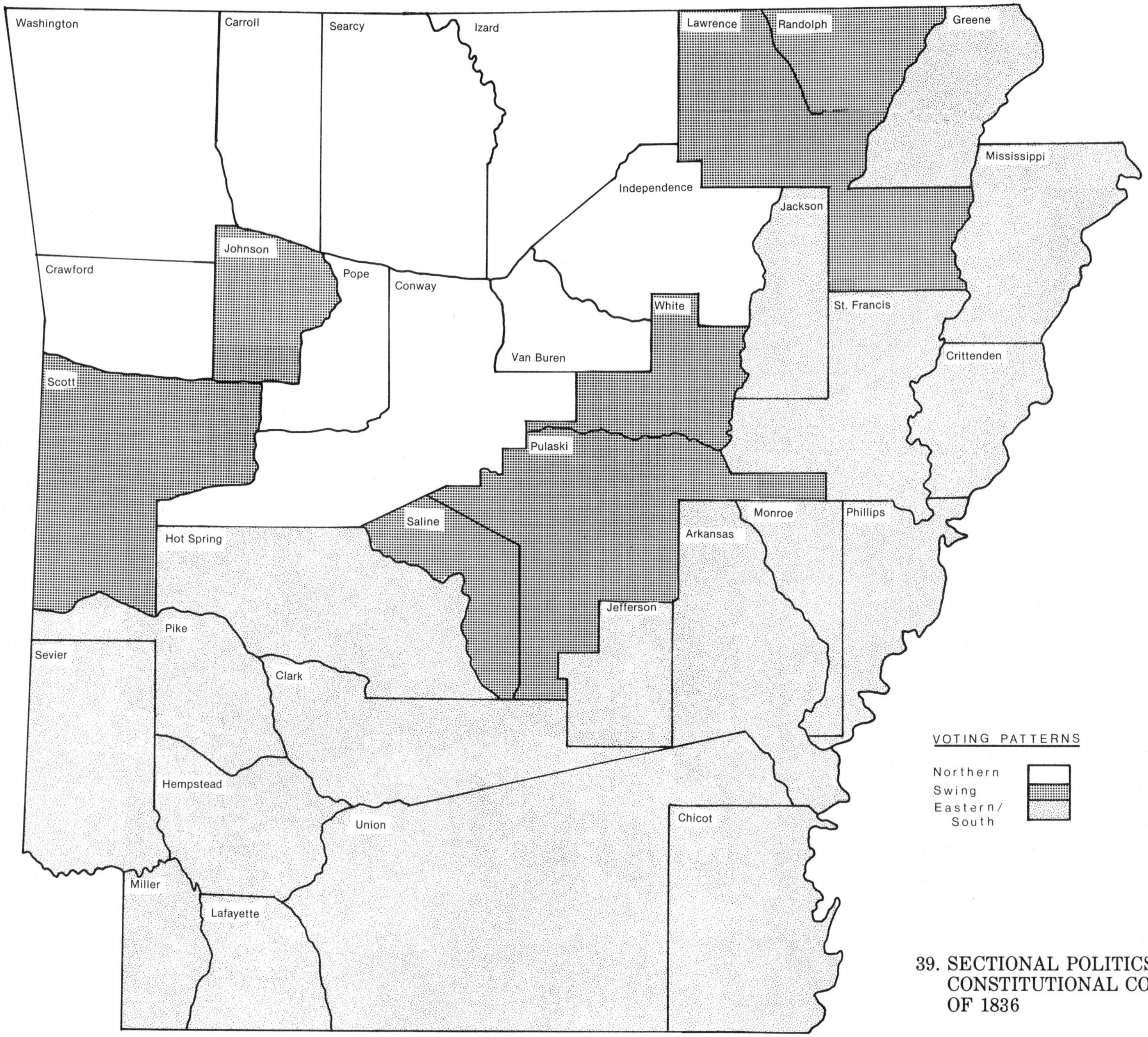

39. SECTIONAL POLITICS IN THE CONSTITUTIONAL CONVENTION OF 1836

39. SECTIONAL POLITICS IN THE CONSTITUTIONAL CONVENTION OF 1836

Since Arkansas first organized politically, geography has been a major cause of political factionalism. The separation of the state into highlands and lowlands with different economies, different societies, and different interests has been the source of repeated struggles. These were apparent during the territorial period and continued after statehood.

At the Constitutional Convention of 1836 the apportionment of the seats in the new state legislature was an issue that quickened sectional hostilities. The issue emerged on January 19, 1836, when the convention's Committee on the Legislative Department delivered a majority and minority report on representation. The majority report had established senate and house districts in such a way that power in the General Assembly rested in the hands of counties in the south and east. The minority report provided for districts that shifted power to the northwest. On January 25 the convention voted twenty-six to twenty-four to substitute the minority report for that of the majority. The substitution spurred a heated debate as delegates from the south and east attempted to overturn the convention's vote. Ultimately, the report was submitted to a special committee, which issued a new report, once again favorable to the south and east. This report provided that the senate districts be altered so that they would contain equal numbers of free white male inhabitants. The house districts were to have five hundred such inhabitants to each representative, with each county entitled to at least one representative. As established, the senate districts were gerrymandered to ensure eastern and southern control. This time the convention adopted the report of the special committee by a vote of twenty-eight to twenty-two.

The votes taken by the constitutional convention clearly showed how the interests of the state were divided geographically. The vote for the Committee on the Legislative Department's minority report on January 25 and the vote against the report of the special committee on January 26 demonstrated solidarity with the position of the northwest. The reverse vote represented support for the position of the south and east. All of the delegates from nine counties and the majority of the delegates from Randolph and Lawrence counties adhered in both votes to the northwestern position. These men represented 49.6 percent of the state's free white males, 24.1 percent of the slaves, and 40.4 percent of the total inhabitants, including women and slaves. All of the delegates from seventeen other counties and the majority from Pulaski County supported the southern and eastern position. These eighteen counties contained 46.0 percent of the free white males, 72.9 percent of the slaves, and 56.1 percent of the total population. The delegates from two counties, Scott and Johnson, western counties that were represented by delegates from outside their boundaries, switched their positions in the two votes.

The south and east won in 1836 and would continue to dominate Arkansas political life during the nineteenth century.

39. SECTIONAL POLITICS IN THE CONSTITUTIONAL CONVENTION OF 1836

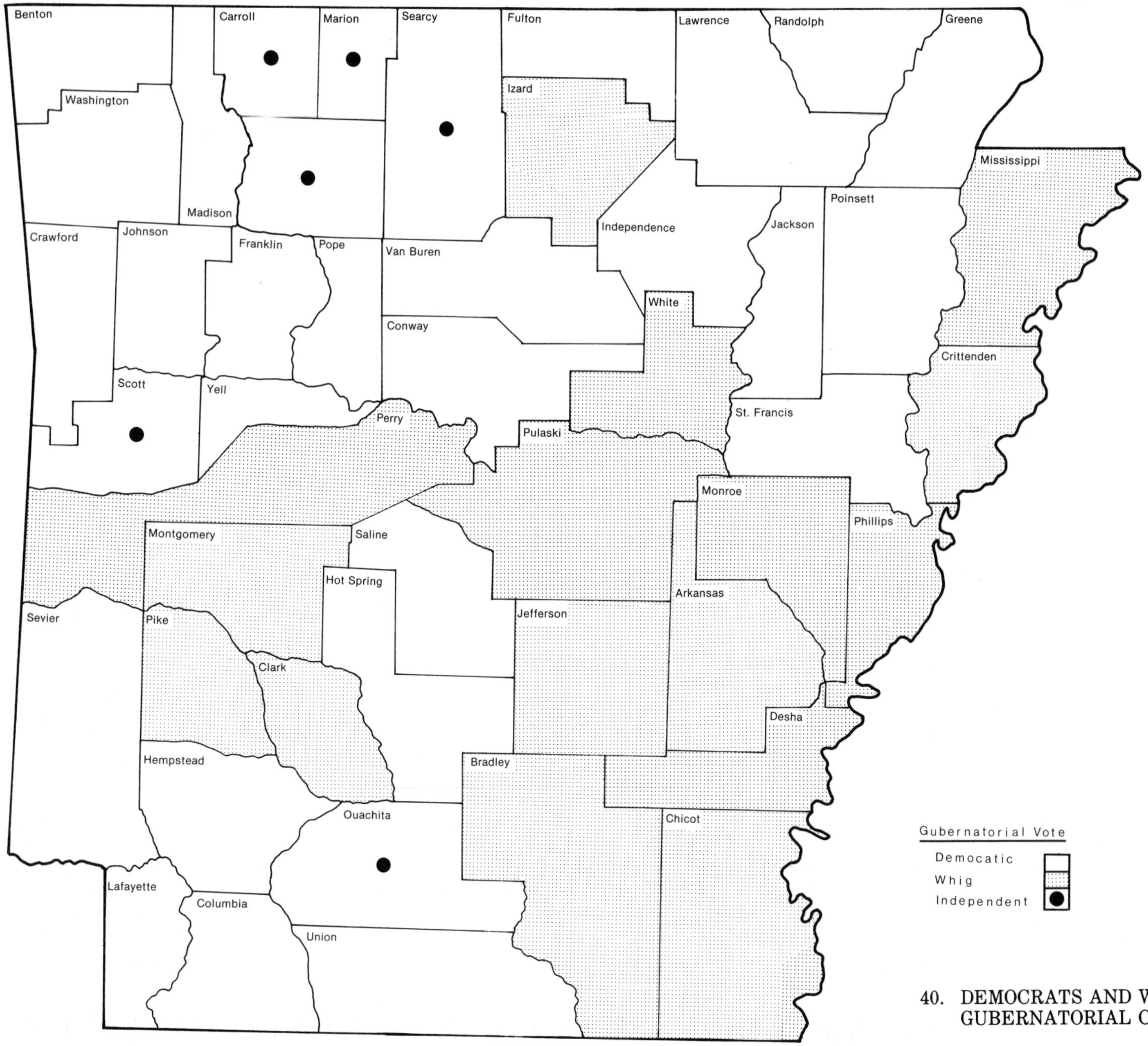

40. DEMOCRATS AND WHIGS IN THE 1844 GUBERNATORIAL CAMPAIGN

40. DEMOCRATS AND WHIGS IN THE 1844 GUBERNATORIAL CAMPAIGN

In nineteenth-century antebellum Arkansas two political parties, the Democratic and the Whig, fought for power. The Democratic party was dominated by a small group of politicians, many of whom were from the interrelated Sevier, Conway, Johnson, and Rector families, called the "Dynasty" or "Family." The Whigs claimed many prominent leaders, such as Albert Pike, Charles F. M. Noland, and David Walker, and were particularly vigorous in the campaigns in the 1840's. Until their collapse between 1850 and 1852, the Whigs were able to maintain such a strong organization that they offered voters a viable alternative, even though the Democrats dominated state government throughout the antebellum period. After 1852 many Whigs moved into the Know-Nothing, or American, party.

Both Democrats and Whigs attempted to secure broad support from among the free white adult male electorate in Arkansas. The rhetoric used by the two parties in appealing to voters varied very little in substance. The Whigs generally expressed belief in a more positive role for government, especially in supporting public improvements such as railroads. Emphasis on the personalities of the candidates, however, was the primary tactic of antebellum campaigning.

Voting patterns do suggest some different interests among Democrats and Whigs. Voting patterns in the 1844 gubernatorial election were particularly interesting because a split within the Democratic party gave the Whigs a chance to elect both the governor and a congressman. In 1844 the leaders of the Democratic party were badly divided, and when a state convention attended by representatives of only sixteen of the state's forty-six counties nominated Elias N. Conway for governor and Daniel J. Chapman for Congress, local groups expressed considerable opposition. The Whigs, in turn, nominated two very popular individuals: Lorenzo Gibson of Hot Springs for governor and David Walker of Washington County for congressman. When it became apparent that the Whigs might win, the Democrats shifted Chapman to the governor's race and persuaded Archibald Yell to run for Congress. Then, as the campaign progressed, Chapman became ill and had to be replaced by a third candidate, Thomas Stevenson Drew.

Despite the Whig threat, the Democrats won. In the gubernatorial race Drew received 8,797 votes, or 47.6 percent of the total. Gibson received 7,066, and Richard C. Byrd, an independant Democrat, secured 2,504. Yell defeated Walker by a vote of 11,112 to 7,576 in the election for the state's only seat in the House of Representatives.

The results of the election demonstrated a pattern that existed through the antebellum period. The Arkansas Whigs, and their successors in the American party, tended to receive their greatest support in those counties which were most urbanized or most closely tied into the market economy. This suggests that in Arkansas the Whigs represented wealthier plantation interests and urban commercial interests. The Democrats, on the other hand, drew support throughout the state, but particularly from the poorer counties.

40. DEMOCRATS AND WHIGS IN THE 1844 GUBERNATORIAL CAMPAIGN

PART SIX
ARKANSAS IN THE CIVIL WAR

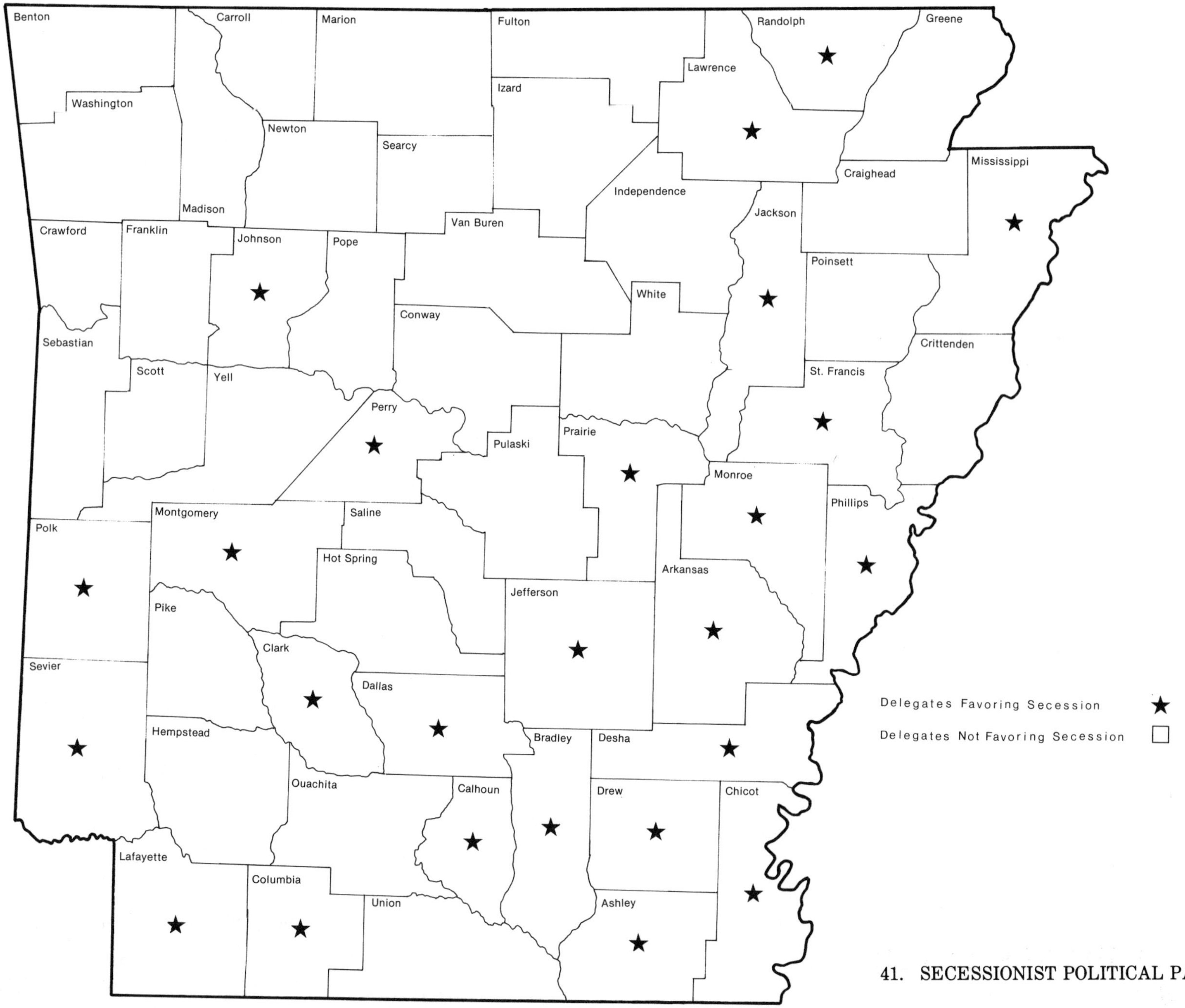

41. SECESSIONIST POLITICAL PATTERNS

41. SECESSIONIST POLITICAL PATTERNS

Arkansas's position during the crisis leading to the Civil War was complex. Its people generally considered themselves to be Southern, but many were reluctant to break with the Union. The state's position on the border of the Indian Territory caused many to fear that a break with the Union would open their settlements up to attacks from Indians loyal to the Union. The still-underdeveloped local system of transportation caused others to hope for federal aid for railroad construction, which certainly was prevented by the Civil War. These problems made people in northern and western Arkansas the strongest supporters of the Union.

Governor Henry M. Rector was a strong proponent of secession during 1860 and 1861, and he pushed the General Assembly to call for the election of a convention to consider the problem of national relations. The legislature ignored the governor's pressure during November and December, 1860. Not until January, 1861, did they pass a bill that called for a referendum on the question of a convention and the selection of delegates. In voting on February 18, 1861, Arkansans decided to hold a convention but elected a majority of delegates to that convention who were against immediate secession. The convention delegates were divided along sectional lines, with the members from the northwest showing the greatest opposition to secession. Yet the Unionism of the northwesterners was not unqualified. Most shared a belief in the right to secede with their secessionist fellow Arkansans. Furthermore, they believed that their state's destiny was bound to the South and its institutions. They did not choose to secede because of Lincoln's election, but they opposed coercion of those states that had already seceded.

On March 4, 1861, the convention met at Little Rock. Despite strong secessionist agitation, the Unionist majority held together and prevented immediate action. They were able to stop the passage of three different measures introduced for the purpose of dissolving the union between Arkansas and the United States. On March 21 the majority forced an adjournment of the convention, although they adopted a resolution that called for a vote of the people on August 5 for secession or cooperation. The convention agreed to meet again on August 19, and if the vote was for secession, they would pass an ordinance of secession. If the vote was for cooperation, the convention would attempt to work with the other unseceded slave states to secure a permanent solution to the sectional conflict. The delegates were concerned that events might move faster than the scheduled vote, however; and they passed another resolution that gave the president of the convention, David Walker, the power to reconvene the session in the event of any emergency.

After the opening of hostilities between the North and the South with the attack upon Fort Sumter, South Carolina, on April 12, 1861, Walker called the convention back to Little Rock. The delegates gathered at the statehouse on May 6. President Lincoln had already called upon Arkansas to furnish volunteers to put down the insurrection in the seceded states. The federal government's call for aid in coercing the states of the deep South made the Arkansas Unionist position untenable. They did not want to secede, yet they were not willing to help President Lincoln put down the secession of the rest of the South. They saw little choice but to join the other southern states. This time the delegates hastily and almost unanimously voted in favor of disunion. Only one of the delegates, Isaac Murphy, from Madison County in the northwest part of the state, refused to vote for secession.

The convention's action made it possible for the state to join the new Confederate States of America. On May 20, Arkansas became a state in the new nation.

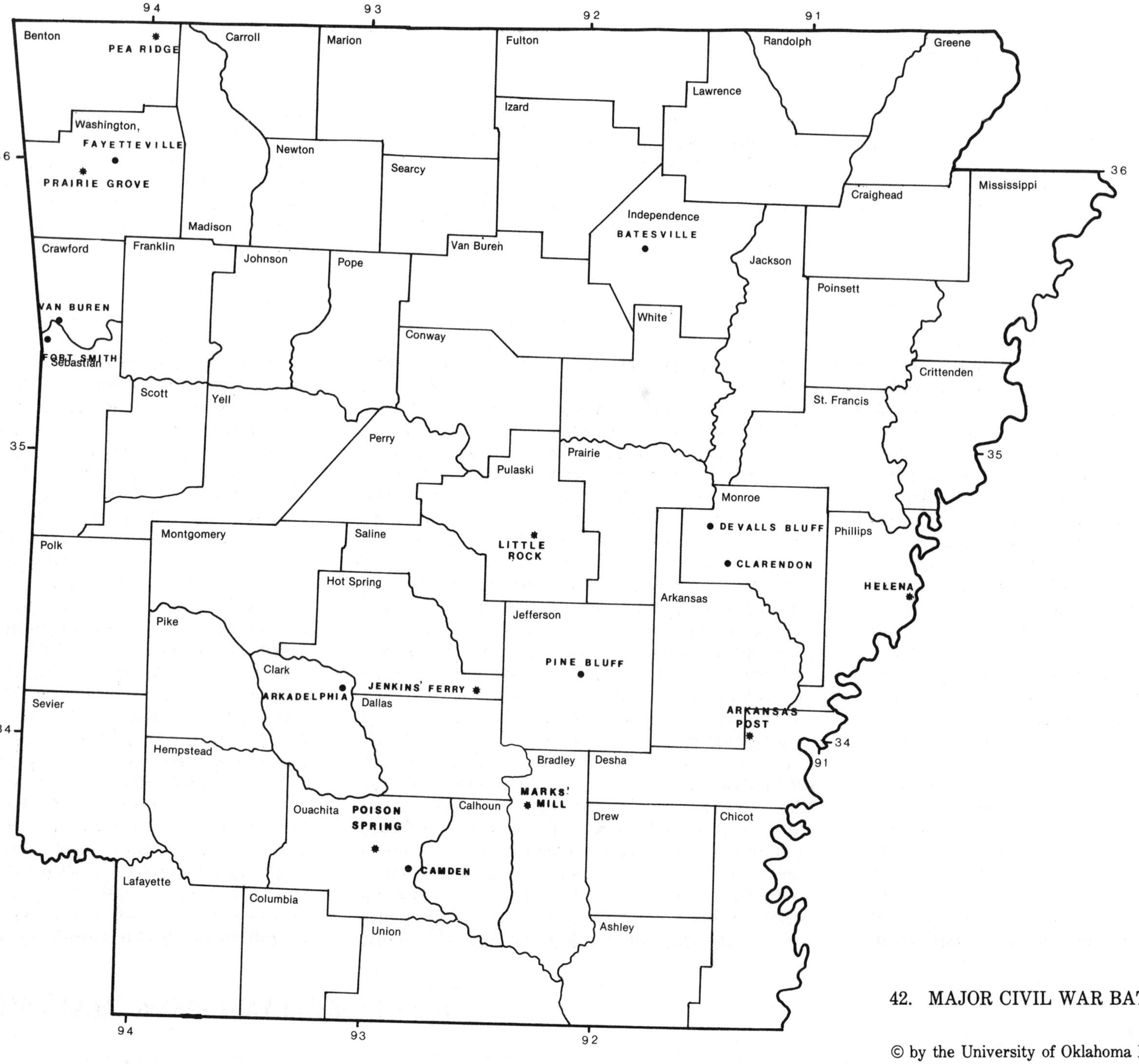

42. MAJOR CIVIL WAR BATTLES

42. MAJOR CIVIL WAR BATTLES

Several major battles occurred in Arkansas during the Civil War. The state's geographical location made it vulnerable to invasion by Union forces from Missouri and along the Mississippi. The Confederate government defended the state because it was essential for the protection of Texas and northwestern Louisiana. Confederate officials worked hard to protect the state during the first part of the war. Their efforts included the negotiation of treaties with various Indian tribes across the western border in Indian Territory. As a result of these pressures, war came to Arkansas early.

The first threat came from Missouri, where a major battle for control of northwest Arkansas took place at Oak Hills, or Wilson's Creek, on August 10, 1861. In that battle the Confederates halted a federal army under General Nathaniel Lyon which was marching against them in southwestern Missouri. Although the battle was a tactical victory for Confederate forces under General Benjamin McCulloch, the Confederates were unable to end the federal threat to Arkansas from the northwest.

In the spring of 1862 another federal army, under the command of General Samuel R. Curtis, pushed Confederate forces out of southwest Missouri and crossed into Arkansas. On March 6, 1862, this movement led to the battle at Pea Ridge, which lasted three days and resulted in the withdrawal of the Confederate army under General Earl Van Dorn from northern Arkansas. This Confederate defeat opened up most of the state north of the Arkansas River to federal military operations. Curtis's army marched down White River in an effort to capture Little Rock, but the failure to establish a line of supply up the White led him to abandon his efforts at capturing the capital, and he turned instead to Helena on the Mississippi. This city fell in June, 1862.

Federal actions in Arkansas were limited after the summer of 1862, as their army's attention turned to the Mississippi. The Confederate commander in Arkansas after May, 1862, General Thomas C. Hindman, took advantage of this diversion to attempt to reestablish Confederate control over northwest Arkansas in an autumn campaign in 1862. Hindman's efforts were fruitless. Federal General James G. Blunt had invaded the Indian Territory and had moved into northwest Arkansas, where he engaged the remaining Confederate forces and planned an attack upon Van Buren and Fort Smith. On December 7, 1862, Hindman's army encountered the federal commands of Blunt and General Francis J. Herron at Prairie Grove. After a battle in which neither side was clearly victorious, Hindman retreated to Fort Smith. His men deserted in droves, and this defeat ended hopes of returning this portion of Arkansas to Confederate control.

In 1863 the federal armies turned their attention to Arkansas once again, and this time their movements led to several major Confederate disasters. On January 8 a federal army began operations against Fort Hindman at Arkansas Post. After only three days, on January 11, that barrier to the use of federal gunboats on the Arkansas and White rivers was removed, and the state was opened to federal penetration. Only federal concern with the capture of Vicksburg, on the Mississippi, delayed an invasion of Arkansas.

To counter the growing federal strength, General Theophilus Holmes, who had assumed command of the Trans-Mississippi Department in the summer of 1862, planned a Confederate counteroffensive in the summer of 1863. His plan called for an attack upon the federal base at Helena. His force was not adequate for that job, however, and was badly mauled in an assault on July 4, 1863.

In August, 1863, General Frederick Steele, just named Union commander at Helena, began to move on Little Rock. Confederate forces were forced to retreat, and the capital was occupied on September 10. The Confederate army withdrew into southwestern Arkansas.

Steele attempted no new operations until the spring of 1864, when he moved against the Confederates in the southwest in cooperation with General Nathaniel Banks's campaign up the Red River in Louisiana. The defeat of Banks at Mansfield, Louisiana, left Steele at Camden, Arkansas, facing an enemy that was quickly concentrating its strength. Confederate cavalry actions at Poison Spring on April 18 and at Marks' Mills on April 25 cut Steele's supply lines and forced him to retreat. A pursuing Confederate army caught up with the retreating Union column at Jenkins' Ferry on the Saline River on April 30, but was unable to prevent Steele's escape to Little Rock.

The battle of Jenkins' Ferry was the last major Civil War engagement in Arkansas. After Steele's return to Little Rock, the Union strategists abandoned efforts at a major offensive in the southwest and pulled off troops for use in the other western armies. Steele and his successor were left to fight a holding action. The Confederates, commanded by General Sterling Price, staged a major raid into Missouri in September, 1864, but they were never able to shake the federal hold over most of Arkansas. Price's defeat led to a degeneration of the war in Arkansas into a small but nasty guerilla operation.

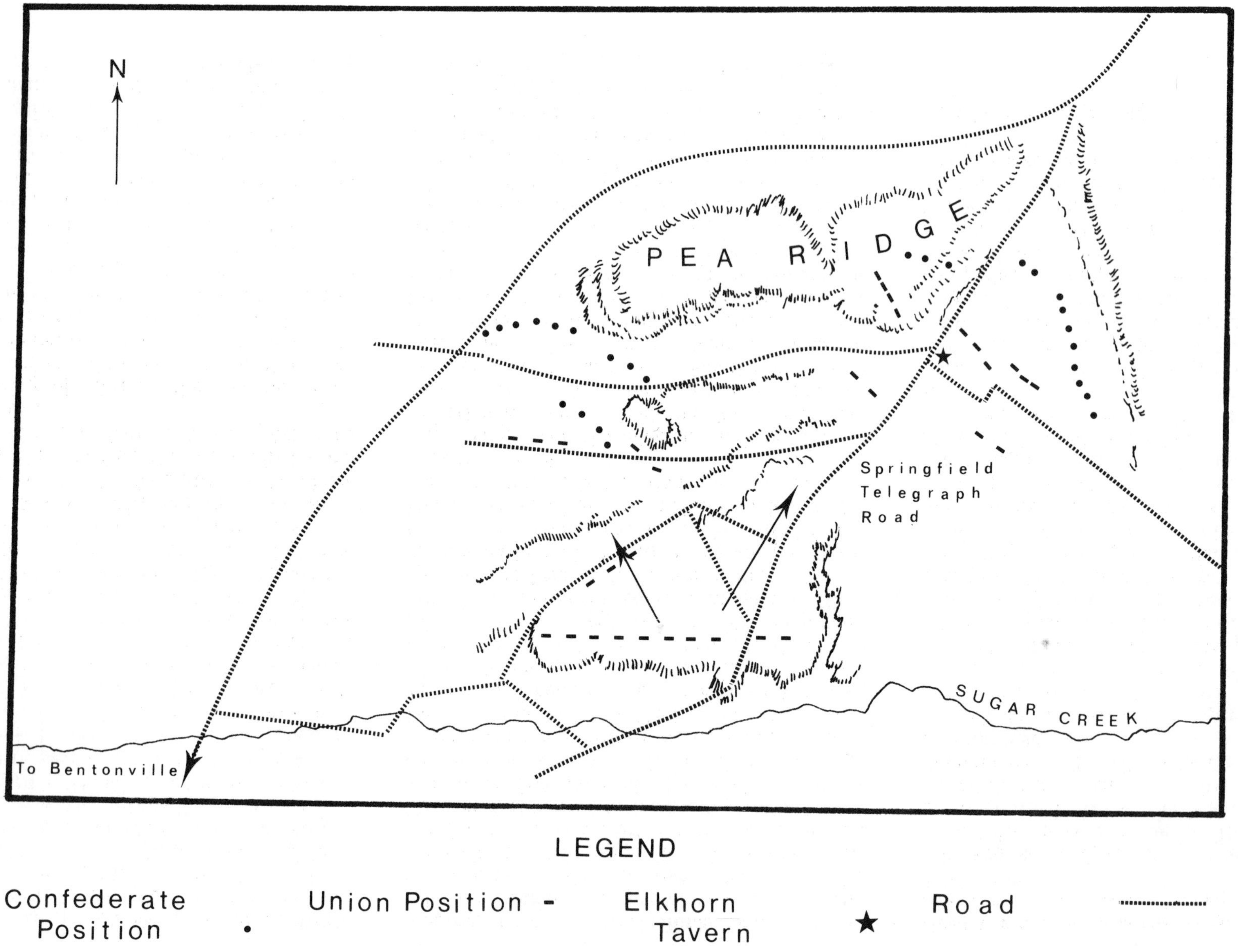

43. CIVIL WAR BATTLE SITES: PEA RIDGE

43. CIVIL WAR BATTLE SITES: PEA RIDGE

In February, 1862, a Union army under General Samuel R. Curtis moved into northwest Arkansas in pursuit of a Confederate force retreating from Springfield, Missouri. Confederate General Earl Van Dorn decided to concentrate his forces in the area to halt the federal drive.

By the first of March, Curtis was on Sugar Creek with a force of about 10,500 men. There he encamped to wait for reinforcements. Van Dorn, who had 16,000 men, decided to attack Curtis before his reinforcements could arrive. He hoped to take advantage of his knowledge of the local roads to surprise Curtis with a turning movement. By moving to the north of the federal force, he thus placed his army between Curtis and the Union commander's line of retreat.

On March 6, 1862, the first engagements began in the battle known as Pea Ridge or the battle of Elkhorn Tavern. On that day Van Dorn attacked a federal force under General Franz Sigel, one of Curtis's division commanders, at Bentonville, southwest of the main federal army, forcing Sigel back onto Curtis's position. Just after sunrise on March 7, Curtis, having figured out Van Dorn's plan of attack, began to reverse his position to face to the northwest. The new line he established was anchored on the north, on Curtis's right, by General Eugene A. Carr's division astride the Springfield-to-Fayetteville Telegraph Road, in the vicinity of Elkhorn Tavern. His left was on Sugar Creek.

On March 7, Van Dorn ordered his right, under General Benjamin McCulloch, to attack Curtis's left. His plan was to pin down Curtis with his initial attack, then throw his own left, under General Sterling Price, into the fight. Van Dorn did not establish a central command post and accompanied Price to the left, thus forfeiting any chance of coordinating his planned attack. McCulloch's attack was underway by midmorning and his forces were heavily engaged for the rest of the day. In the fight McCulloch and General James McIntosh were both killed. At about 2:00 P.M. the attack of the Confederate right slackened and Van Dorn threw his left against Curtis in strength. Curtis had figured out Van Dorn's plan, however, and had begun reinforcing his right before the attack began. Curtis's reinforcements began to arrive on the field to find Union forces under Colonel E. A. Carr engaged in a desperate fight. Although they had been pushed backward, they had not broken, however; and the arrival of reinforcements at about 5:00 P.M. stopped Van Dorn's attack.

On March 8, Curtis redeployed his army to face in strength Van Dorn's concentration, forming a line directly across the Telegraph Road. The federal left quickly extended itself onto Pea Ridge, thus exposing the Confederates to a flanking fire. Curtis's right then moved forward to put the Confederates into the center of a concentrated fire that began to break Van Dorn's line. The Confederate commander, short on ammunition, decided he had no choice but to withdraw, and at ten o'clock he ordered his army to begin a retreat southeast towards Huntsville.

The battle was a costly affair. Over 26,000 men fought on the two sides. Curtis lost 203 killed, 972 wounded, and 176 missing. Van Dorn lost 185 killed, 525 wounded, and 300 missing. After the battle Van Dorn led his army from Arkansas to Mississippi, leaving the state almost defenseless. The battle ended all chances for the Confederacy to take Missouri, capture Saint Louis, or otherwise threaten the communication lines of General William Halleck's and General Ulysses S. Grant's campaign in western Tennessee and northern Mississippi.

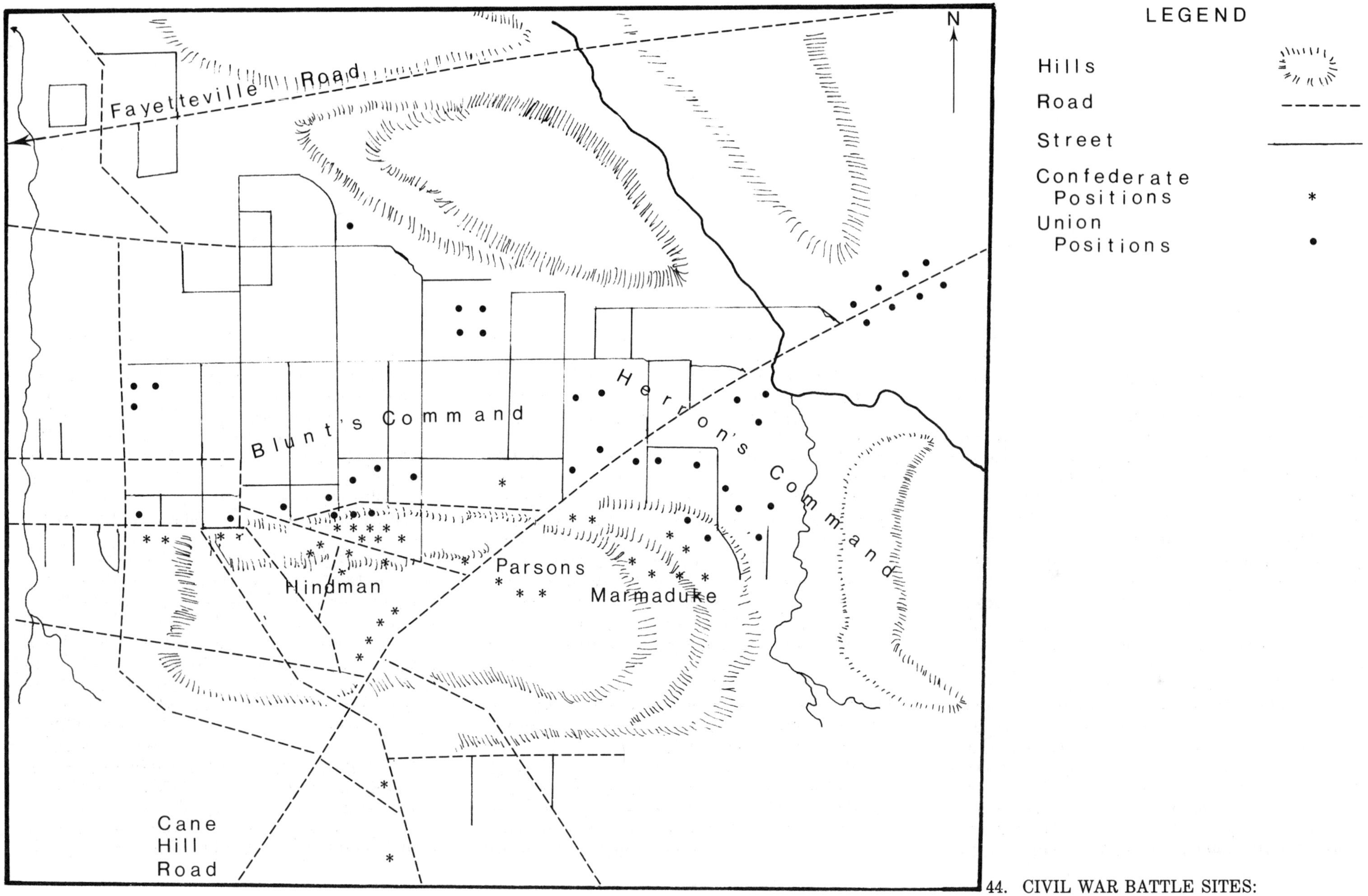

44. CIVIL WAR BATTLE SITES: PRAIRIE GROVE

44. CIVIL WAR BATTLE SITES: PRAIRIE GROVE

Confederate leaders refused to give up northwest Arkansas. In 1862, General Thomas Hindman attempted to lift federal pressure on Fort Smith with an attack against Union armies in the Cane Hill area near Fayetteville. Hindman put together a force of nine thousand infantry, two thousand cavalry, and twenty-two pieces of artillery and marched to the north. His original plan was to attack the command of General James G. Blunt, about eight thousand men, at Cane Hill. Cavalry skirmishes took place in the vicinity of the Union position on December 5 and 6. On the evening of December 6, Hindman discovered that a relief column of as many as six thousand infantry, five hundred cavalry, and thirty guns, under the command of General Francis J. Herron, was almost at Fayetteville. Hindman decided that on the next day he would get between Blunt and Herron before their forces could join, smash Herron on the northeast, then turn his full force on Blunt. He could thus use his superior force against the two smaller forces rather than having to face a larger force if they united. On December 7, 1862, Confederate and Union forces fought a major battle at Prairie Grove.

Marching out at 4:00 A.M. on December 7, Hindman ran into Herron's cavalry along the Cane Hill Road. The Union cavalry retreated until stopped by the main body of Herron's infantry about six miles from Fayetteville. Hindman occupied a position astride the road along a wooded hill overlooking a prairie to the north. Rather than attack, however, Hindman prepared to receive an assault by Herron and thus lost precious time. At about 1:00 P.M. Herron's forces moved against Hindman's right. Herron's infantry mounted two frontal assaults on the Confederates, which were met by shot-guns, rifles, and muskets. Each assault was repelled.

Hindman's delay allowed Blunt to arrive at the scene of the battle while Herron engaged one of the Confederate divisions under John S. Marmaduke. Blunt did not have his troops in position to attack until after Herron had been driven back. Another frontal attack aimed at a line held by a Confederate brigade under General Mosby M. Parsons ended with the federals being repelled. After a cannonade directed at the Confederate line, Hindman's troops faced a third charge along their entire front. Once again they beat back the federal attackers. As the day ended, the federals mounted a fourth unsuccessful assault.

At the end of the day on December 7 both sides held their original positions, but Hindman had failed to prevent the joining of Blunt and Herron's commands, which now outnumbered him. Furthermore, Union cavalry were beginning to operate at his rear and threaten his supply train and line of retreat. In addition, he had sustained heavy losses—164 dead, 817 wounded, and 336 missing. Blunt and Herron had lost 175 killed, 813 wounded, and 263 missing, but these had not been enough to alter the balance of strength, which was in favor of the Union forces. If all this was not enough, Hindman was also low on ammunition, food, and forage. He concluded that he had no option but to retreat. Although he claimed a victory, Hindman's force had clearly suffered a loss. The federals still controlled northwest Arkansas and were poised for an attack on the Confederate positions at Van Buren and Fort Smith.

44. CIVIL WAR BATTLE SITES:
PRAIRIE GROVE

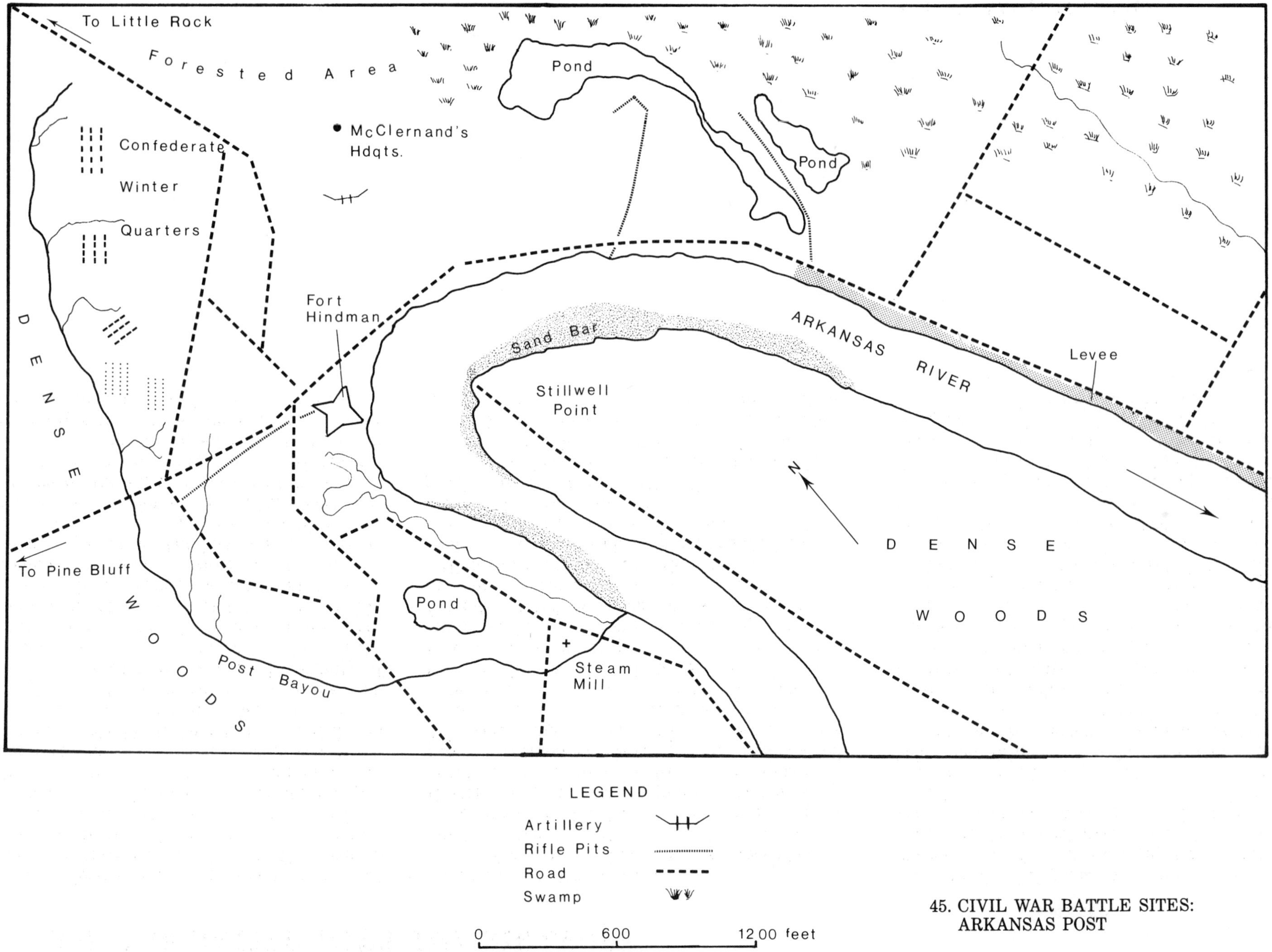

45. CIVIL WAR BATTLE SITES: ARKANSAS POST

45. CIVIL WAR BATTLE SITES: ARKANSAS POST

By the winter of 1862–63 the federal forces in the west were engaged in opening the Mississippi River to the Gulf. Vicksburg was the last major post south of Memphis, but a Confederate position at Arkansas Post had been used to harass federal communications along the Mississippi, firing on transports and even seizing an occasional vessel. The main Confederate force was at Fort Hindman, which was near Arkansas Post and just above the conjunction of the Arkansas River with what was known as the White River cutoff. Fort Hindman lay on the north bank of the Arkansas on a bend in the river. About a mile downriver from the fort were extensive trenches serving as the Confederate's first defensive line. Commanded by General Thomas J. Churchill, a garrison of some five thousand men was positioned there.

To remove the obstacle presented by the Confederate command at Arkansas Post was the object of an expedition led by General John A. McClernand in January, 1863. The federals mustered an army of thirty-two thousand men, supported by naval forces. McClernand's command steamed up White River, took the cutoff, then moved up the Arkansas and landed three miles below Fort Hindman at 5:00 P.M. on January 9, 1863. The troops unloaded through the night and were not disembarked completely until noon on January 10.

On the morning of January 10, while McClernand's army was still unloading, the Union gunboats moved closer to the Confederate position and shelled the fort and the trenches. By 11:00 A.M., McClernand had his first columns moving against the position in an effort to flank Churchill. By the middle of the afternoon Churchill had withdrawn his entire force to the fort and an inner line of trenches. Although the federals pressed forward, they made no concerted attack, and Churchill's men worked at strengthening their position. In the evening, however, the federal gunboats again moved in, and they fired on the fort for two hours. That evening Churchill received a telegram from the Confederate commander in Arkansas, General Theophilus Holmes, ordering him to "hold out till help arrived or until all dead."

The main attack on Fort Hindman began on January 11 with a simultaneous navy and infantry assault. The gunboats fired for three hours, destroying all of Churchill's guns except one, and two of the federal gunboats passed the fort in order to place the Confederate position in a crossfire. While the gunboats threw shell at the Confederates, the infantry charged headlong at the works. Although they repulsed many assaults, the Confederates were outnumbered and undersupplied. The ultimate conclusion of the battle was inevitable, and in the middle of the afternoon the garrison surrendered, although Churchill denied that he had ordered a white flag.

The federal forces captured the entire garrison. McClernand reported that he had taken 5,000 prisoners, as well as animals and large quantities of arms and ammunition. Some Confederates effected an escape, but not enough to keep the loss from being a major one for the South. Churchill reported that he had lost 60 killed and 80 wounded. The federals had lost 134 killed, 898 wounded, and 29 missing.

The battle of Arkansas Post removed the force at the rear of Grant's operations against Vicksburg. More critical for Arkansans, it destroyed another major Confederate army positioned to protect the state. The loss of Churchill's garrison left General Theophilus Holmes with few troops to prevent a federal invasion of Arkansas from the Mississippi. Only Grant's preoccupation with the seizure of Vicksburg delayed such an operation.

45. CIVIL WAR BATTLE SITES:
ARKANSAS POST

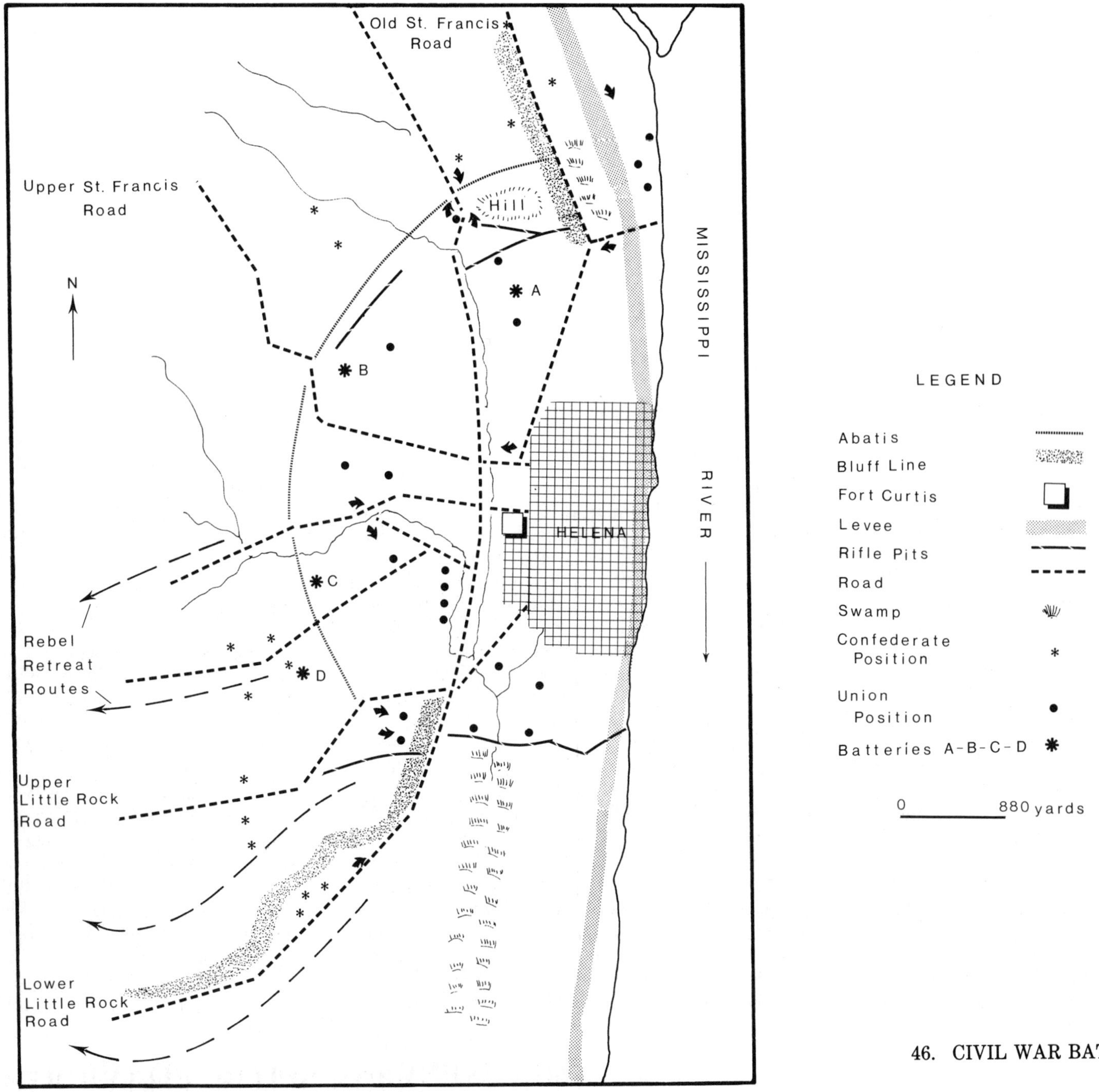

46. CIVIL WAR BATTLE SITES: HELENA

46. CIVIL WAR BATTLE SITES: HELENA

Helena had been occupied by Union General Samuel R. Curtis in June, 1862, and while the federals had shown no inclination to invade Arkansas, their base at Helena was a major thorn in the side of Confederate Arkansas. From there the federals staged cotton raids that picked up not only cotton but large numbers of slaves. Furthermore, Helena was a major base used by Ulysses S. Grant as he pushed towards Vicksburg. In part to relieve pressure on Vicksburg, and also to remove a great "advantage to the enemy," General Theophilus Holmes planned an attack on Helena that came on July 4, 1863.

The federals had been at Helena for a year by the summer of 1863 and had turned it into a formidable position. General Benjamin M. Prentiss had a force there of 4,129 officers and men, well dug in. Just west of the town was a major earthwork fortification named Fort Curtis, which contained siege guns with a range of at least one mile that could fire on the western approaches to the town. Along a ridge running parallel to the Mississippi west of Helena, the federals had placed a series of smaller forts known as batteries A, B, C, and D. Each of these positions was supported by entrenchments. To make an approach even more difficult, trees had been felled on the western side of the ridge to impede an attacking force and give the federal cannon a clear field of fire. The Confederate attacking columns faced a difficult task.

Holmes attacked Helena with 7,640 men. He staged his movement using three columns. General John S. Marmaduke led Holmes's left, General Sterling Price was at the center, and General James F. Fagan was on the right. The main attack was to be made by Marmaduke, but a federal flanking operation pinned him down. The burden of the assault then fell upon Fagan and Price. Although the Confederates were supposed to attack at daylight, the battle began with an attack by Fagan against Battery D and with Price not yet in position. By the time Price attacked Battery C, Fagan's assault had been blunted. Price's men charged with bayonets and drove the federals from their position. Holmes rode into the captured battery and attempted to move his men to the support of Fagan with a flanking attack on Battery D on the south. On the point of victory, the Confederate forces fell apart in the confusion of the battle. Rather than taking Battery D, the relief force was caught in a murderous fire from the position's defenders and from the guns of Fort Curtis and the gunboat *Tyler* offshore.

Unable to regain control over his troops, Holmes ordered a general retreat at 10:30 A.M. After about five hours of battle the Confederates had been forced to withdraw. In his report the general announced that he had lost 173 killed, 687 wounded, and 776 missing. Many of the missing had been captured in the attack on the Battery D. The federal force lost 57 killed, 146 wounded, and 36 captured or missing. Confederate failure at Helena for all practical purposes demoralized the Confederate army in Arkansas. Holmes was forced to move his army all the way back to Little Rock. His infantry would never again penetrate the eastern part of the state.

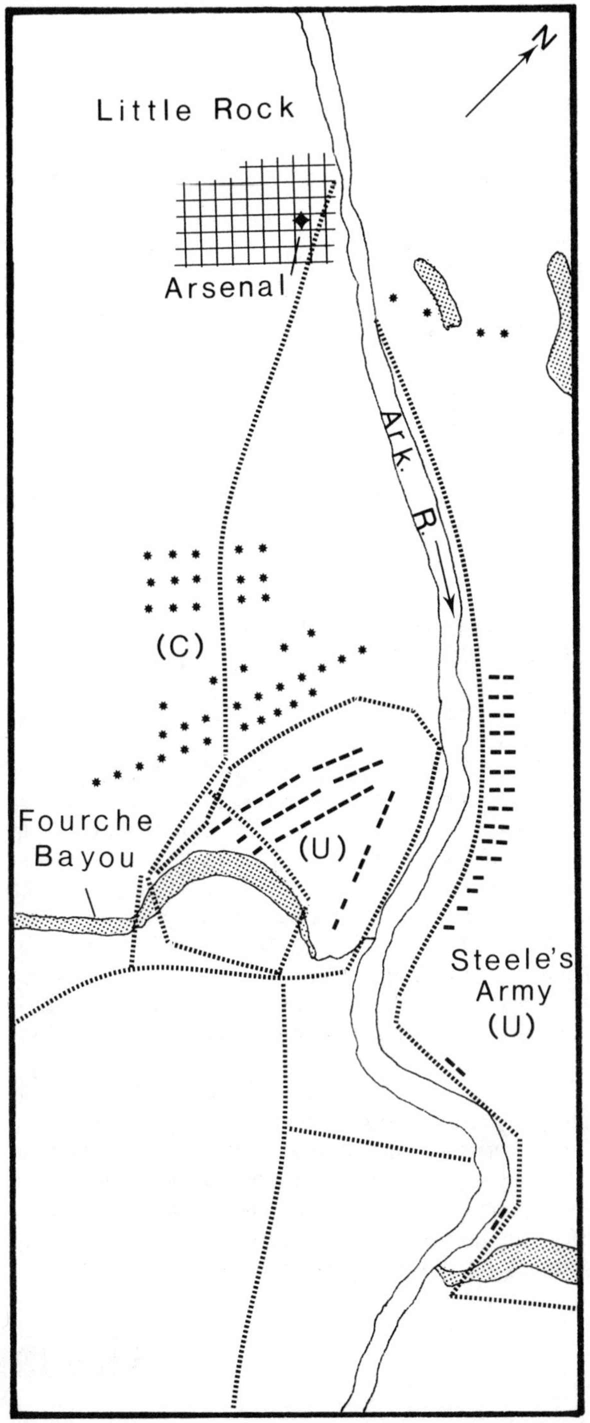

47. CIVIL WAR BATTLE SITES: LITTLE ROCK

47. CIVIL WAR BATTLE SITES: LITTLE ROCK

The inevitable invasion of the interior of Arkansas began in August, 1863, when Union General Fredrick Steele advanced with some thirteen thousand men from Helena towards Little Rock. His route took him first to Clarendon, then to De Valls Bluff, where he began to move overland towards Little Rock. At De Valls Bluff he established a critical supply base where riverboats could deposit the materials necessary for his short overland move to the state capital.

The Confederate command was in disarray. On July 23, General Theophilus Holmes, because of his ill health, ordered General Sterling Price to assume command in the field. Price had about twelve thousand men to stop Steele, although he estimated only eight thousand were effectives. Price hastily prepared for Steele's coming by constructing defensive works on the north bank of the Arkansas River near Little Rock. By the latter part of August he had concluded that Steele intended to approach Little Rock along the southern bank of the Arkansas, and on August 31 he ordered General L. Marsh Walker to concentrate Confederate cavalry on the south bank east of the city, though most of the infantry were left north of the river. On September 7 at Ashley's Mill, north of the river and about eight miles below Little Rock, federal cavalry drove away a defending force of Confederates to open the way for a crossing of the river. On September 9, Price ordered the construction of a defensive line on the southeast side of Little Rock.

Steele's main army marched up to the river at Ashley's Mill on September 9. Steele's cavalry crossed the river and made possible the laying of a pontoon bridge upon which he could cross his infantry. Resistance by Colonel Archibald S. Dobbins was quickly brushed aside. On June 10, Steele moved his cavalry and two regiments to the south side of the Arkansas and began to move upon Little Rock in two columns, one on each side of the river. Price was forced to throw all of his cavalry to the south side of the river to hold up Steele's advance, but this could only be a holding operation. The Union column on the south of the river ran into resistance on Fourche Bayou, but flanking fire from federal artillery on the north bank of the Arkansas drove the Confederates from that position.

At 11:00 A.M. on June 10, Price ordered the withdrawal of his infantry and artillery from the positions taken on the north side of the river and began pulling out of Little Rock, destroying the supplies and equipment that the troops could not carry with them. Steele's crossing of the river made the defense of Little Rock impossible, and as he moved up both banks of the river, he presented Price with no choice but to begin a complete evacuation of the town. By 5:00 P.M., Price had all of his forces out of the city and on the road towards Arkadelphia, where he regrouped and attempted to establish another defensive line.

The capture of Little Rock was a brilliant operation on the part of General Steele. In a campaign that had lasted forty days from beginning to end, he had lost 18 men killed, 118 wounded, and only 1 missing. Price gave no report of his losses, but Steele wrote that he estimated Confederate losses at nearly 1,000 men, mostly prisoners captured as his army moved into the city. The loss of Little Rock further dampened the morale of the Confederate army in Arkansas, marking the third major defeat in 1863. Hundreds of men left the ranks of Price's army as it headed for Arkadelphia. Only the failure of Steele to pursue prevented what might have been the total collapse of organized Confederate resistance in Arkansas.

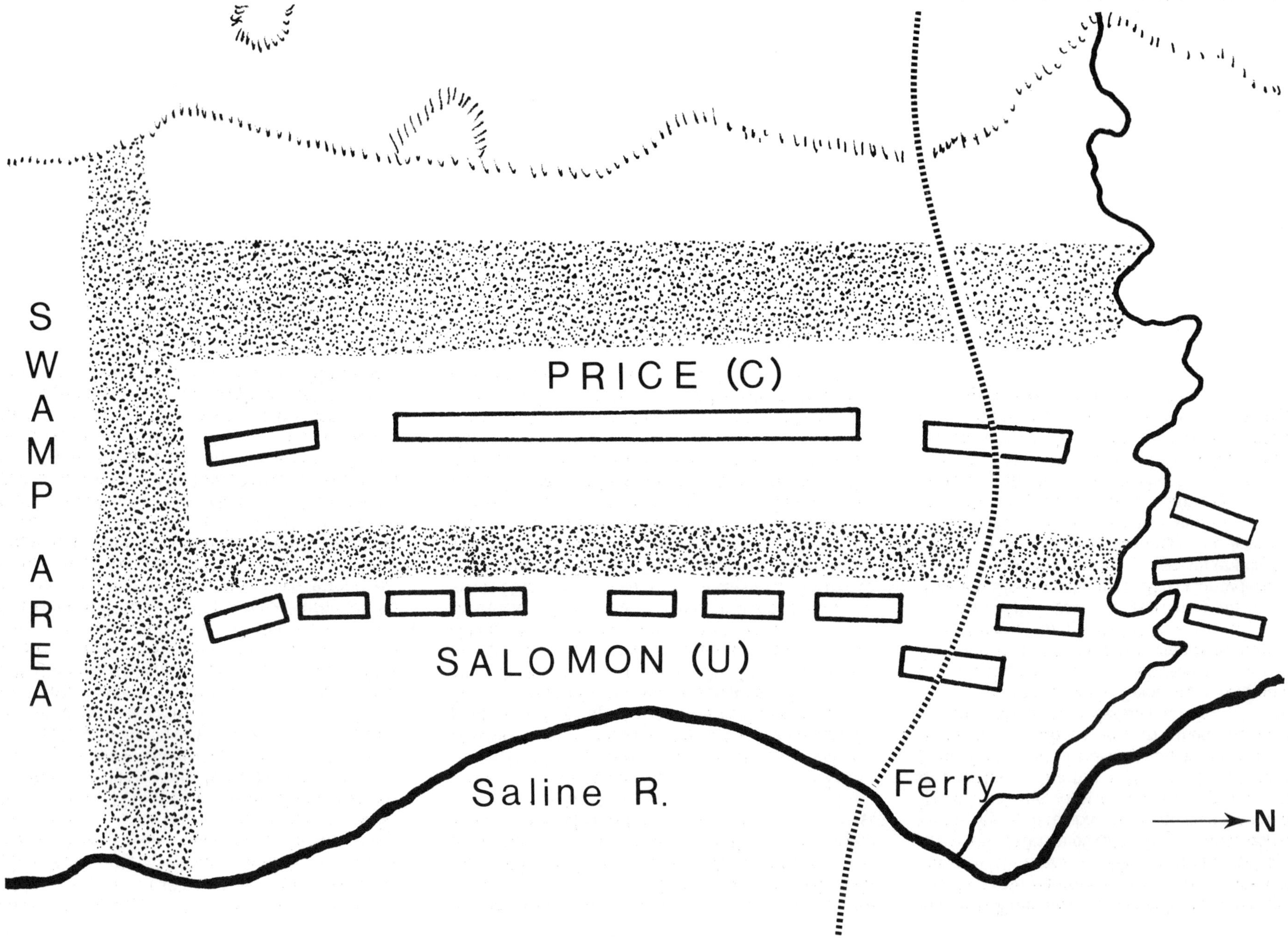

48. CIVIL WAR BATTLE SITES: JENKINS' FERRY

48. CIVIL WAR BATTLE SITES: JENKINS' FERRY

On March 23, 1864, a federal army commanded by General Frederick Steele marched out of Little Rock towards the southwest to link up with General John M. Thayer's Frontier Division in an operation that was coordinated with General Nathaniel Banks's Red River expedition to drive Confederate forces out of western Arkansas and Louisiana and eastern Texas. Steele's command, after combining with Thayer's, included 13,754 officers and men. The Confederate army in Arkansas, under the command of General Sterling Price, was outmatched and fell back, awaiting reinforcements from the Confederate army at Shreveport, Louisiana. In battles outside of Shreveport, Banks was defeated, and Confederate General E. Kirby Smith sped large reinforcements of infantry to the aid of Price.

Steele suddenly found himself with an exposed supply line, deep in enemy territory, with little hope of reinforcement. At first he moved his Union troops into Camden on the Ouachita River in hopes of securing provisions over that river, but the move did not produce the desired results. In a battle at Poison Spring on April 18 and another at Marks' Mills on April 25 he was cut off from supplies and was forced to abandon Camden before becoming completely surrounded.

Steele's retreat from Camden set up the major battle that was fought at Jenkins' Ferry on April 30, 1864. Jenkins' Ferry, a crossing on the Saline River, was the last spot that Confederate pursuers had a chance of catching up with the fleeing federals, who had thrown a pontoon bridge across the stream and were drawn up for a crossing there. At daylight April 30, General Sterling Price's calvary, under John S. Marmaduke, engaged Steele's rearguard. Price immediately ordered General Thomas J. Churchill's Arkansas infantry division to support Marmaduke and held another division under General Mosby Monroe Parsons in reserve. The battlefield provided limited room for maneuver, and Churchill deployed his troops in a line that had no choice but a frontal assault upon the federals. The force they were fighting was General Frederick Salomon's infantry division, which Steele had ordered to protect the crossing. Salomon's line was protected on his right by an impassable bayou, and on his left by a heavily wooded swamp, and his reserves were in a position to be thrown to any point threatened. Price sent Parsons's division as well as that of General John G. Walker into the battle, but Salomon's line did not budge. After an assault at 10:30 A.M. the Confederates did not renew the attack, and Salomon's men withdrew across the Saline. After crossing, the federals destroyed their bridge and much of their equipment and moved back to Little Rock, where they arrived on May 2.

The battle at Jenkins' Ferry was the last major engagement fought in Arkansas. Steele's army would remain in Little Rock, protecting a line from Fort Smith to the Mississippi River. Price's army would regroup for an attack into Missouri in September, 1864. The two armies would never fight each other again, however.

48. CIVIL WAR BATTLE SITES:
JENKINS' FERRY

PART SEVEN
ARKANSAS AND THE NEW SOUTH, 1865–1950

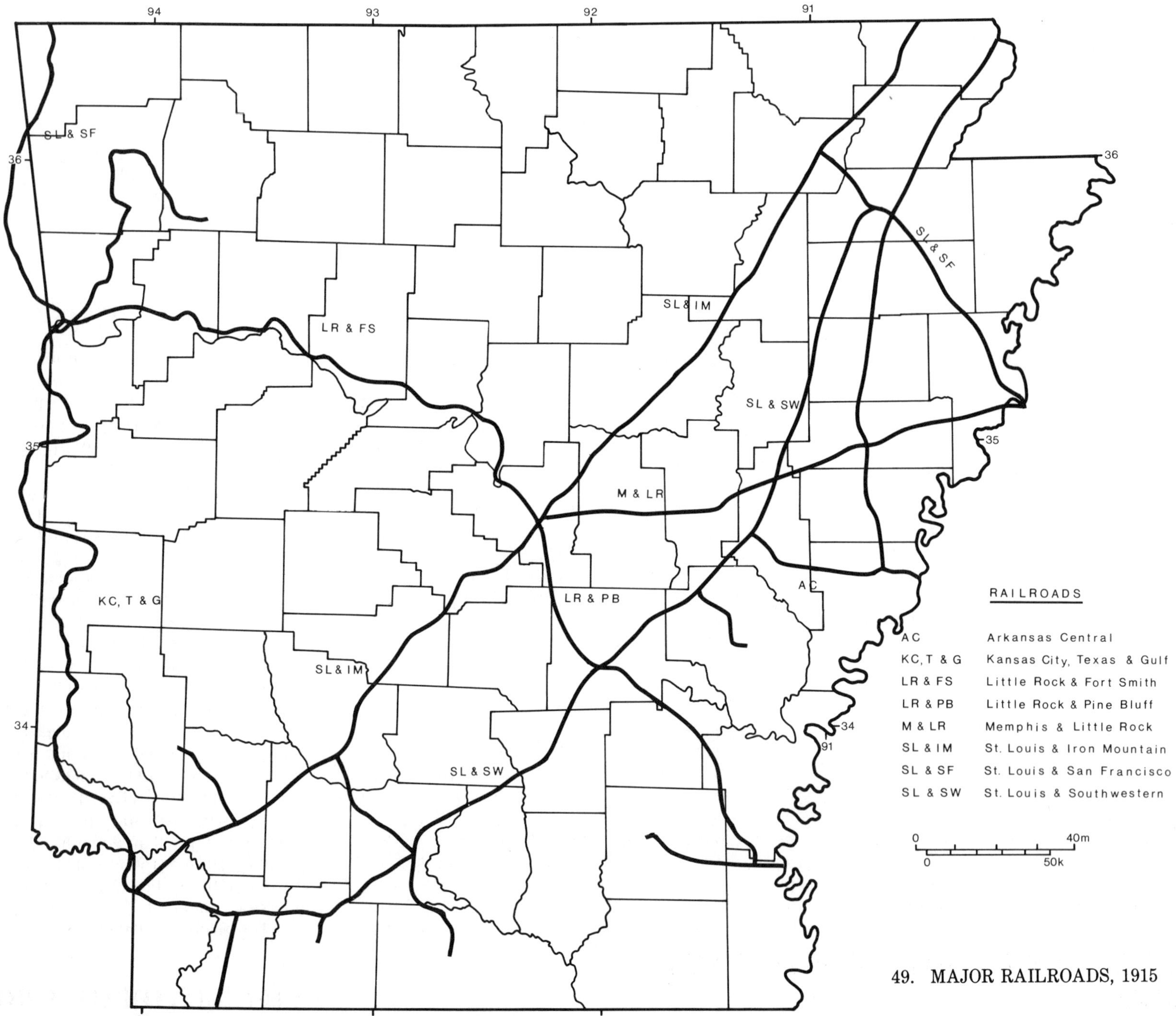

49. MAJOR RAILROADS, 1915

49. MAJOR RAILROADS, 1915

After the Civil War, Arkansans witnessed the introduction of the railroads, a new means of transportation that revolutionized life in the state. The new lines overrode many of the state's natural barriers and tied Arkansans closer together than ever before. With the railroads also came changes in the economic life of the state's people.

Before 1860, Arkansans had chartered many railroad companies, and though early development was slow, these first companies set out the basic network of main lines that would later be built in the state. The first company to operate trains in Arkansas was the Memphis & Little Rock, chartered in 1853 for the purpose of connecting those two cities. Track was laid from Hopefield to the Saint Francis River by 1858, and four years later a second link was completed from De Valls Bluff on the White River to Little Rock. The gap between the White and the Saint Francis was completed after the Civil War, under the supervision of Nathan Bedford Forrest. In addition to the M & LR, the legislature chartered Roswell Beebe's Cairo & Fulton in 1853 to run from southeast Missouri to Fulton on the Red River; and the Little Rock & Fort Smith, to run west through the Arkansas River valley, received a charter in 1854.

The actual construction of these lines boomed after the Civil War. With government aid in the form of land grants and cash subsidies, all of the lines envisioned in the late 1850's were finished. The Memphis & Little Rock was completed on April 11, 1871. By 1874 the Cairo & Fulton, as an extension of the St. Louis, Iron Mountain & Southern, had completed tracks from the Arkansas's northeastern border to the southwest and trains were running from Saint Louis to Texarkana. In 1874 the Little Rock & Fort Smith completed its line to the Oklahoma border. At the end of the 1870's various roads had completed 822 miles of track.

The laying of track continued after 1880. In 1882 construction began on the St, Louis & Texas that ran from Clay County in northeast Arkansas to Pine Bluff to Texarkana. It was reorganized in 1891 into the St. Louis & Southwestern, or "Cotton Belt." Another major line that developed during this period was the St. Louis & San Francisco, which ran northwest from Memphis through Crittenden, Craighead, Poinsett, Lawrence, Sharp, and Fulton counties. A branch of this road ran along the northwestern border of the state from Benton County to Fort Smith. A parallel line to the St. Louis & San Francisco branch was the Kansas City, Texas & Gulf, or Kansas City Southern, which ran across Benton County, then down the Oklahoma border, entering Arkansas in Scott County and continuing on the Arkansas side of the border to Texarkana.

Consolidation also took place. The M & LR combined with the Choctaw, Oklahoma & Gulf in 1898 to form a line from Memphis to Oklahoma City. By the end of the century 3,167 miles of track ran through the state, of which 2,383 were main lines and 784 consisted of branches or spurs. The basic pattern of the state's railroad system is shown in this 1915 map.

Railroad construction peaked in the early 1930's. The major trend was consolidation. By 1932 five trunk-line railroads operated 4,022 miles of track in the state, over 81 percent of the total. The largest was the Missouri Pacific, with 1,830 miles, including the old LR & FS and the Cairo & Fulton. The St. Louis & San Francisco, usually known as the Frisco, operated 727 miles. The Chicago, Rock Island & Pacific, including tracks of the M & LR and the Choctaw, Oklahoma & Gulf, was the third-largest road, with 703 miles of track. The St. Louis & Southwestern owned 608 miles, and the Kansas City Southern operated 154 miles.

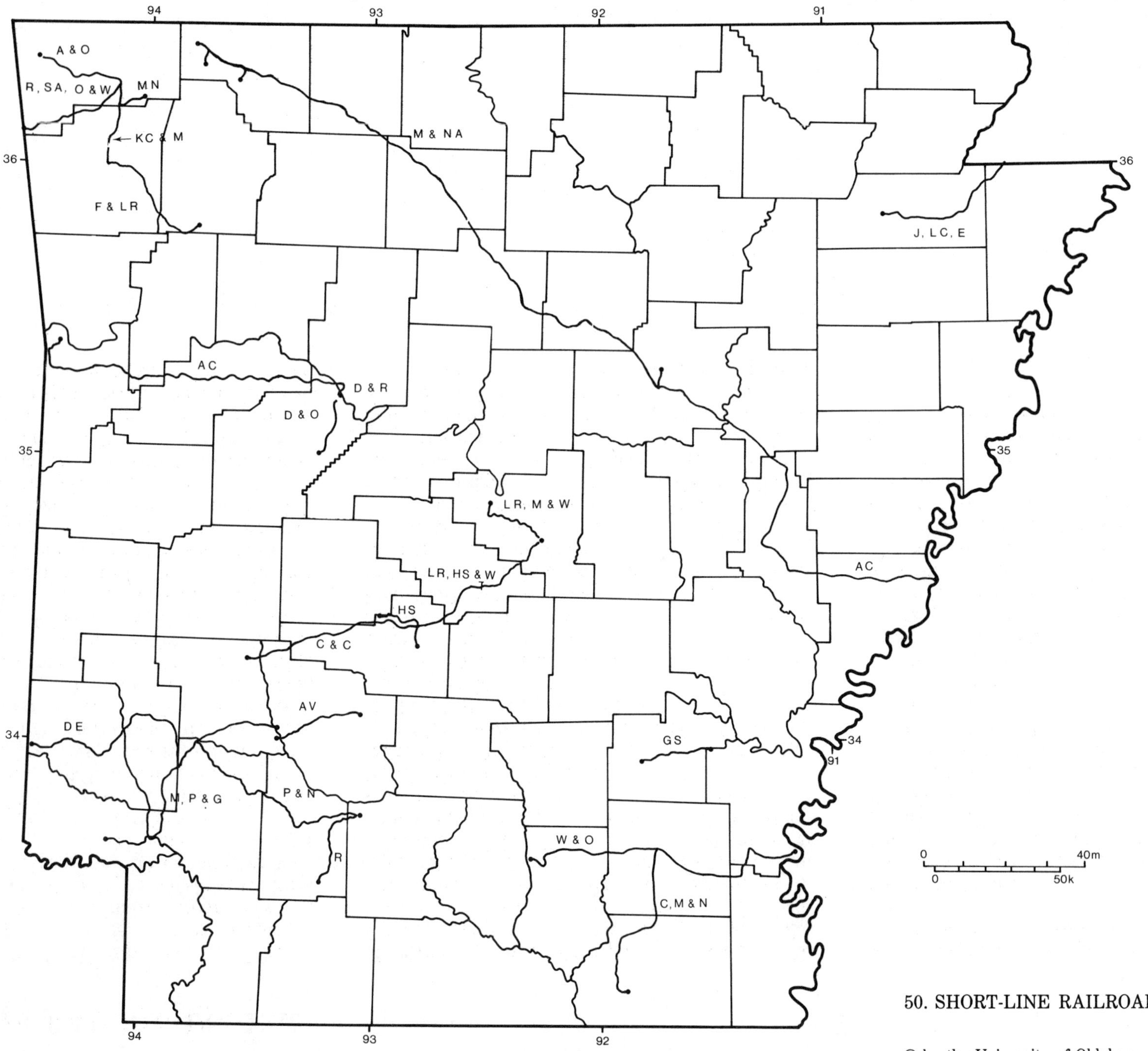

50. SHORT-LINE RAILROADS, 1876–1914

50. SHORT-LINE RAILROADS, 1876–1914

An important aspect of railroad construction in Arkansas was the proliferation of many short-line roads that were not a part of the major trunk-line systems. Generally, these roads developed because local interests sought to tie their town or county to the state and national transportation system after being bypassed by one of the bigger roads. Backed first by local support and then by whatever state aid could be obtained, such railroads were built throughout Arkansas. Many of these lines were integrated into the larger lines in the 1880's and 1890's, but others remained independant operations throughout their existence. The short-line roads connected the countryside with the great trunk lines, feeding freight and passengers into the national systems. This map shows some of the more important short lines.

In 1900 there were thirty-three short-line roads in operation. The average length of these roads was twenty-eight miles; the shortest line possessed less than a mile of track, the longest had seventy-four. Two decades later the mileage of the short-line roads peaked. The number of roads continued to grow in the first two decades of the century, and in 1931–32 the Arkansas Railroad Commission reported forty-five in existence, but by then the average number of miles run by these roads had decreased to twenty-one miles.

There was no typical line. Rather, each met the needs of a particular area or developer. The Hot Springs Railroad serves as a good example. It was developed in 1875 by Joseph ("Diamond Joe") Reynolds of Chicago, who built the road himself so that he could use the baths of Hot Springs without having to ride a stagecoach to the spa. The road was built in three-foot narrow gauge and ran between Hot Springs and Malvern, where it connected with the St. Louis, Iron Mountain & Southern. The first train ran over the line on January 25, 1876. The twenty-three-mile-long Hot Springs Railroad continued to operate as Reynolds's private road through the rest of the century.

While constituting only a small fraction of the total railroad mileage in Arkansas, these short-line roads played a major part in developing the state, hauling freight and passengers before the emergence of a modern highway system in the twentieth century.

Short-Line Railroads of Arkansas

AC	Arkansas Central	1898
AN	Arkansas Northwest	1898
A & O	Arkansas & Oklahoma	1898
AV	Antoine Valley	1907
C & C	Caddo Choctaw	1901
C,M & N	Crossett, Monticello & Northern	1911
DE	DeQueen, Eastern	1905
D & O	Dardanelle & Ola	1908
D & R	Dardanelle & Russellville	1883
F & LR	Fayetteville & Little Rock	1887
GS	Gould, Southeastern	1909
HS	Hot Springs	1876
J, LC, E	Jonesboro, Lake City, Eastern	1908
KC & M	Kansas City & Memphis	1914
LR, HS, W	Little Rock, Hot Springs, Western	1876
LR, M & W	Little Rock, Maumelle, & Western	?
M, P & G	Memphis, Paris & Gulf	1907
MN	Monte Ne	1902
M & NA	Missouri & Northern Arkansas	1909
P & N	Prescott & North-western	1891
R	Reader	?
R, SA, O & W	Rogers, Southeastern, Arkansas, Oklahoma, & Western	1898
W & O	Warren & Ouachita	1898

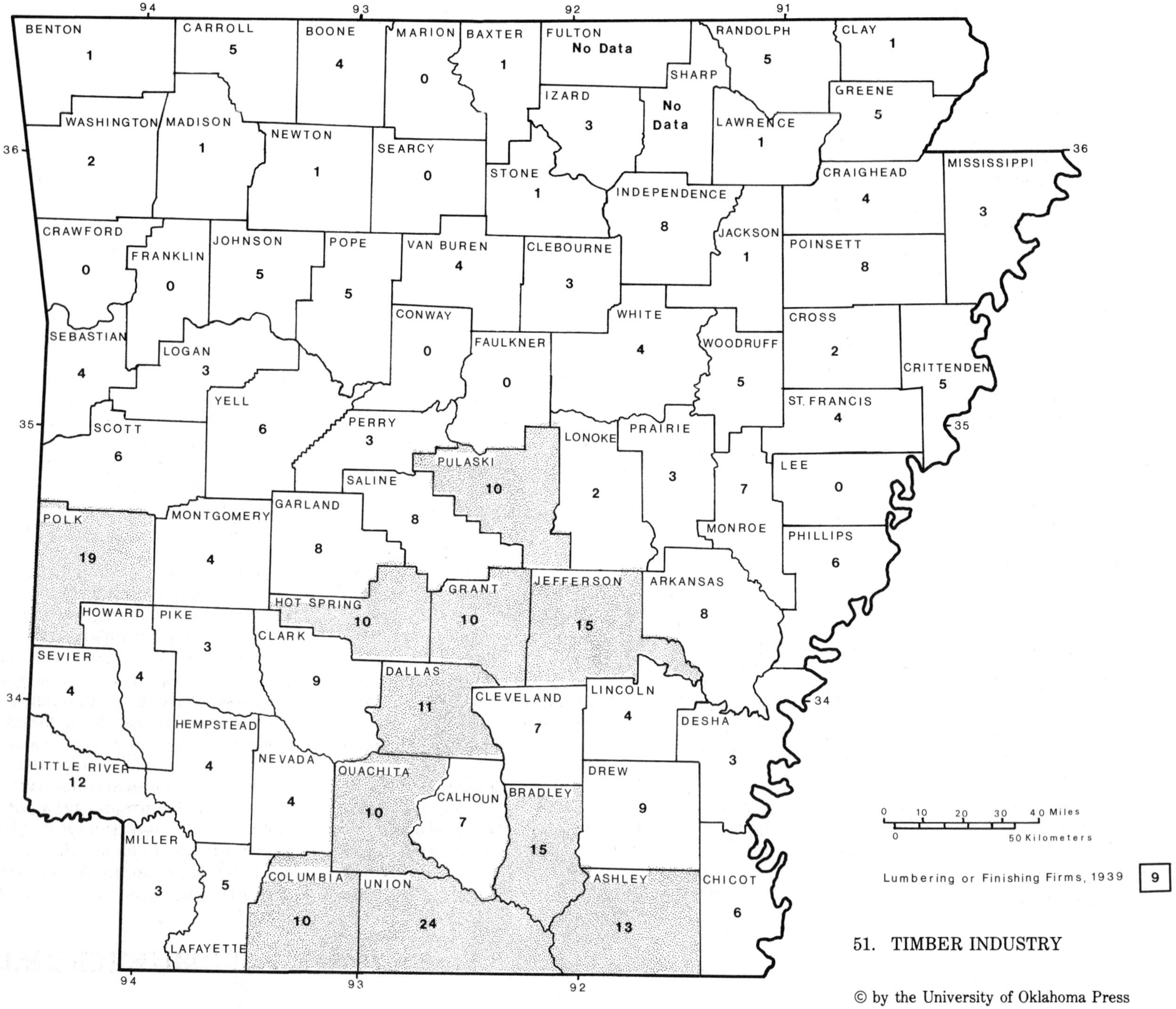

51. TIMBER INDUSTRY

51. TIMBER INDUSTRY

Among the important economic developments made possible by the expansion of railroads in Arkansas was the emergence of the state's timber industry. As shown in Map 8, stands of hard and softwood forests grow throughout the state. At the beginning of the twentieth century these forests covered over three-fourths of the total area. The railroads made access to them possible and encouraged the exploitation of this resource.

The cutting of the state's forests began in earnest in the 1880's, encouraged by the railroad developers. As rails were laid in the forests, construction camps became logging camps, and sawmills were built at many of these points. The state also witnessed a rapid expansion of mills and factories engaged in making products from the cut timber—mills to make railroad crossties, shingles, boxes, barrel staves, doors, and even furniture. By 1909, Arkansas was the fifth-largest producer of lumber in the nation, and related industries employed over half of the state's nonfarm work force.

The industry was destroying itself, however. Until 1920 the logging and lumber campanies carried out a cut-and-run business. They denuded the countryside of the best timber and then moved on. Such practices almost destroyed the forests. The best timbers were gone by the 1920's, and the industry faced hard times.

At that time, however, some companies began to change their ways, emphasizing conservation of the existing forests and redevelopment of the cutover areas. The shift allowed the timber industry to continue as one of Arkansas's most important economic activities. In 1977, 18 million acres of forest land were still being worked in the state, and the industry produced almost 49 billion board feet of lumber.

Statistics on the lumber industry are difficult to secure. Map 51 shows the number of establishments in each county engaged in lumbering or a timber-products industry in 1939, at a time when operations had recovered from the lows of the 1920's. The map indicates the extensive nature of the industry and its major foci.

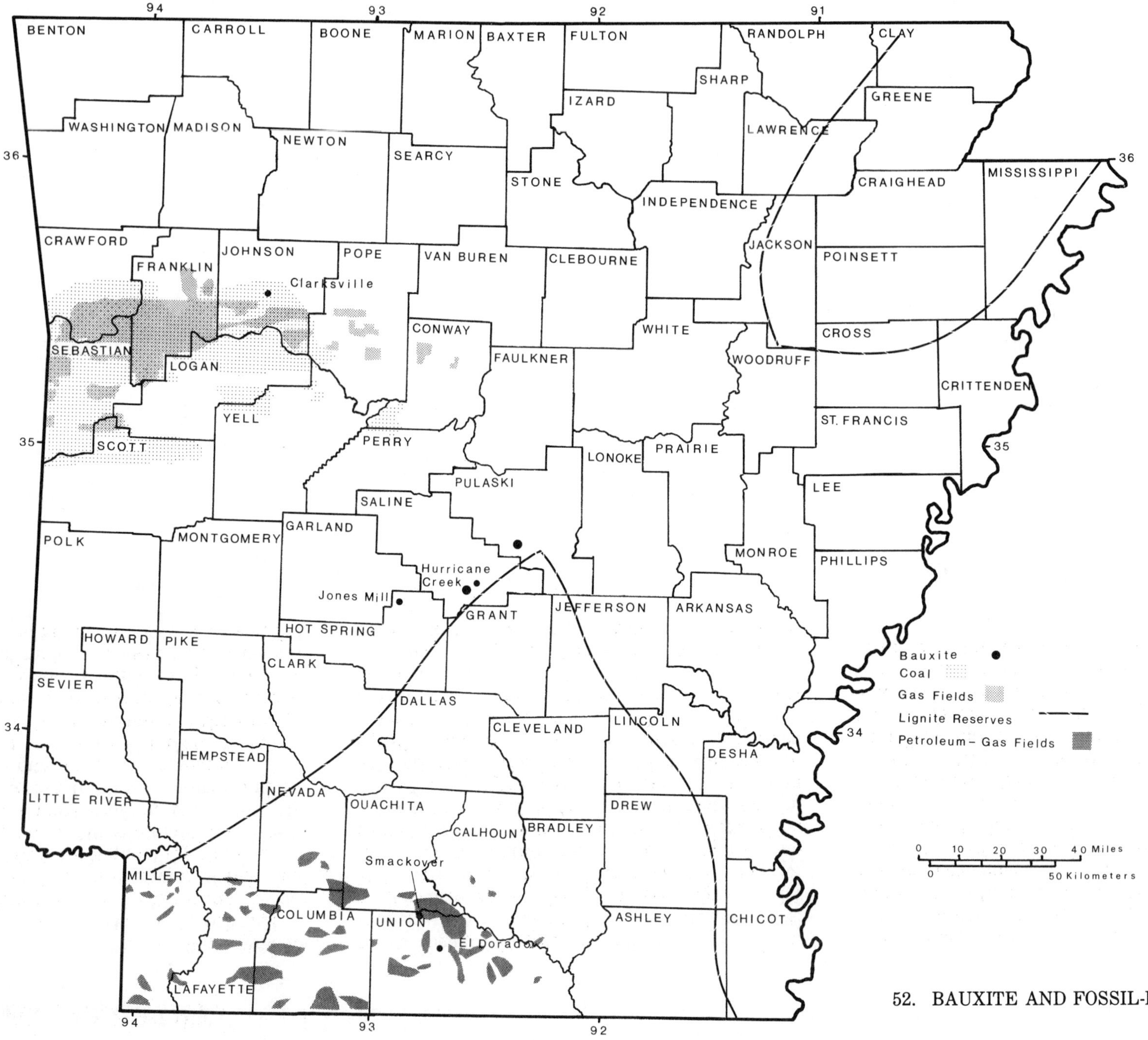

52. BAUXITE AND FOSSIL-FUEL MINING

52. BAUXITE AND FOSSIL-FUEL MINING

In addition to spurring the timber industry, railroad expansion in Arkansas spurred development of major mineral resources. The most important mineral products in Arkansas, economically, have been fossil fuel and bauxite.

Bauxite, an ore out of which aluminum is made, is found primarily in Pulaski and Saline counties. Bauxite mining began in 1899, although knowledge of deposits preexisted actual exploitation. The industry expanded rapidly. By the 1920's, Arkansas was producing nearly 95 percent of the nation's bauxite. Much of the best ore was dug out during this period. During World War II the federal government built two major processing plants, one at Hurricane Creek and the second at Jones Mill. After the war these operations were turned over to private industry. Bauxite has continued to be an important mine product of the state. The playing out of deposits and foreign competition have led to reduction in mining activities and the closing down of plants, yet in 1969 bauxite reserves in Arkansas were estimated at 100 million tons, leaving the state as a major source of future production. In 1980, Arkansas produced 1,299,000 metric tons of the mineral, although since then mining has been reduced significantly.

There are major deposits of bituminous coal and some semianthracite in the Arkansas River valley. The two most important coal producers in the state are Johnson and Sebastian counties. The first coal mined in the state was in Johnson County in the 1840's, but more extensive mining was not possible until the completion of the Little Rock & Fort Smith Railroad opened the deposits to the northern market in 1883. During this period private contractors worked the mines with convict laborers. By 1920 there were 140 coal operations in the state, working deposits in Sebastian, Franklin, Johnson, Pope, Logan, Scott, Crawford, Washington, and Yell counties. Until 1920 this mineral was the most valuable of the state's mineral products. Although there was a brief flurry of activity in the coal fields in the midst of the energy crisis, present-day operations are relatively small.

Across the entire West Gulf Coastal Plain there are deposits of petroleum, but most of Arkansas's petroleum comes from two counties, Columbia and Union. In 1920 the Hunter Oil Company brought in the first producing oil well in Ouachita County. This was followed by more discoveries near El Dorado in Union County. The Smackover field, in the northern part of Union County, was the largest, reaching peak production of 77,398,000 barrels of oil in 1925. All of these fields have continued productivity.

Connected with the state's coal and petroleum deposits are major natural gas pockets. One such field in the area of Crawford, Scott, Franklin, and Sebastian counties, has been the most important commercial natural-gas source in the state. The first gas well was drilled there in 1901, and in 1920 it produced an average of 200 million cubic feet of gas per day. Other major sources of gas have been located in the southern Arkansas oil fields and near Clarksville in Johnson County.

Value of Arkansas Bauxite, Coal, Oil, and Gas
(in millions of dollars)

Year	Bauxite	Bituminous Coal	Oil and Natural Gas
1880		0.03	
1890		0.30	
1900	0.01	2.50	
1909	*	3.50	0.10
1919	*	5.30	0.60
1929	*	6.10	
1954	13.10	2.80	70.10
1958	16.50	2.10	80.30
1972	22.50	3.40	79.50
1977	*	15.30	283.30
1982	18.90	5.20	957.30

*No information provided by census because only one company was in operation.

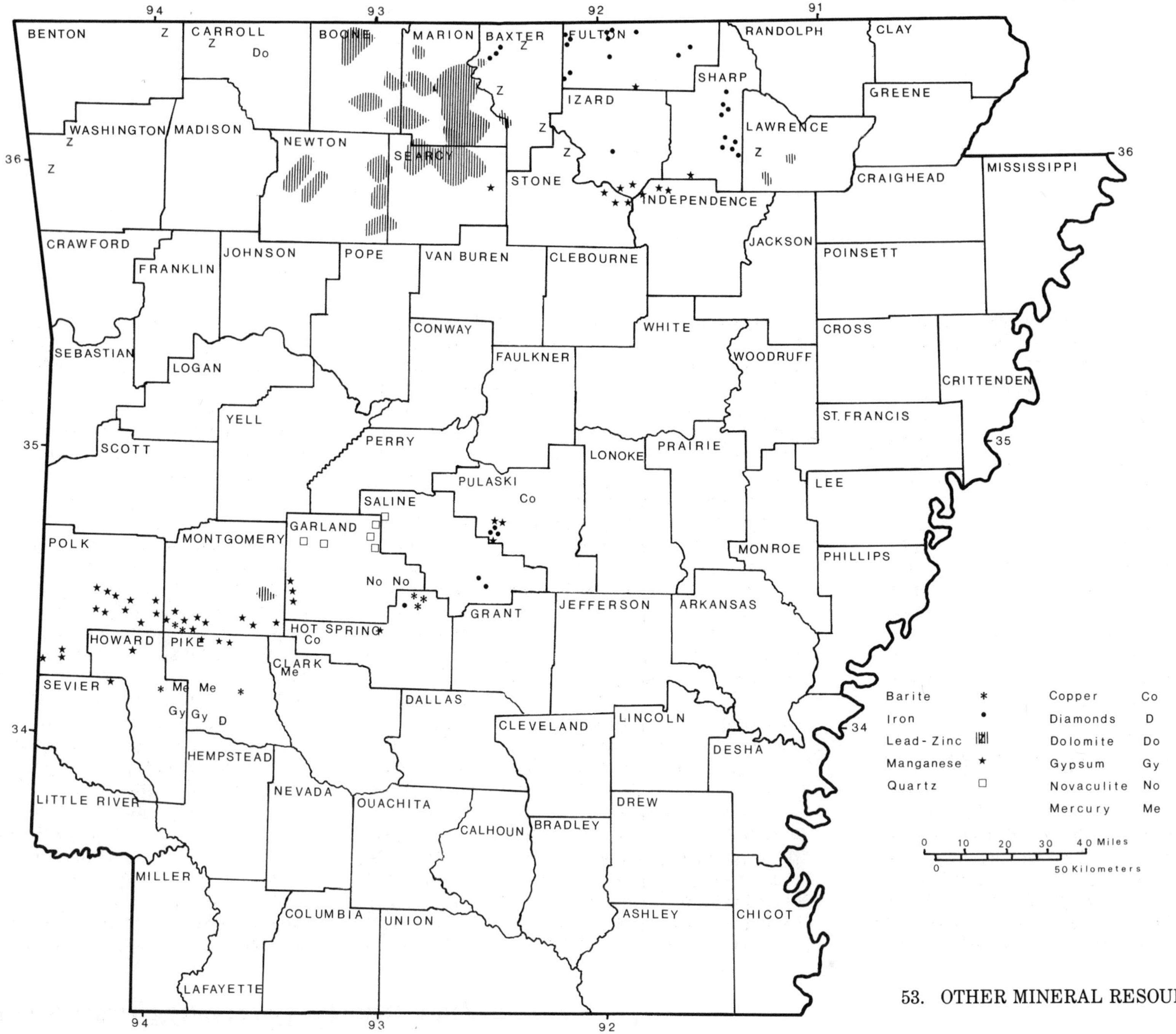

53. OTHER MINERAL RESOURCES

53. OTHER MINERAL RESOURCES

In addition to its important bauxite and fossil-fuel deposits, Arkansas possesses many other mineral resources to which Arkansans historically have looked to generate income. The quality and quantity of the deposits, the expense of mining, and problems of transportation, however, have prevented the development of the vast majority of these resources.

In 1984 the U.S. Bureau of Mines reported the most important resourses in the state, other than bauxite and the fossil fuels, to be stone, valued at $59,800,000; sand and gravel, worth $29,993,000; and clays, valued at $7,838,000. Small amounts of abrasives, bromine, chalk, lime, and tripoli were also mined in that year. Except for bromine, which is extracted from brine taken from the oil fields in south Arkansas (see Map 52), all of these minerals are widespread across the state. The accompanying map shows that the other economically viable minerals have been found in more restricted locations.

Lead and zinc mining early was considered economically viable. Extensive deposits are located in Baxter, Benton, Boone, Carroll, Lawrence, Marion, Montgomery, Newton, Searcy, Stone, and Washington counties in the north. Traces (not shown) exist in Garland, Hot Spring, Howard, Pike, Polk, Pulaski, Saline, and Sevier counties in the central part of the state. Development began in 1857, with active mining after 1875, but inadequate transportation restricted growth. Although commercial mining continued in Marion County until 1962, most of the state's lead and zinc mines closed after World War I.

Copper deposits are situated in the Ouachita and Ozark mountains and also in Hot Spring and Pulaski counties, but in small concentrations.

Diamonds are found in Pike County, but efforts to commercially mine them have not been successful and there has been no commercial mining since 1962.

Iron ore is deposited in Baxter, Fulton, Izard, Hot Spring, Saline, Pulaski, and Sharp counties. The quality of the ore, however, has not justified development, although there has been some sporadic production. Reserves of over 120 million tons makes iron ore a mineral with mining potential in the future.

Major manganese deposits are located in Sevier County, with smaller deposits in Independence, Izard, Stone, Garland, Hot Spring, Montgomery, Pike, Polk, Pulaski, and Saline counties. Small operations worked these deposits until 1959.

Barite, used in drilling mud, was found in Hot Spring, Howard, Montgomery, and Pike counties, but commercial production ended in 1982.

Mercury has been extracted from cinnabar in Howard, Pike, and Clark counties. Mining flourished during World War II, but has not been active since then.

Novaculite, used for whetstones, is found in Garland County. Deposits of gypsum, a mineral used for plaster, Portland cement, wallboards, and fertilizers, are located in Howard and Pike counties. Important deposits of dolomite, used in ceramic materials, are in Carroll County.

Other minerals, not shown on this map, exist in small deposits, but have some economic potential. These include stibnite, found and produced in Sevier County until 1947; molybedenite and titanium in Hot Spring County; nickel in Saline County, mined prior to the 1880's; phosphate rocks in Independence, Izard, Searcy, Stone, and Van Buren counties; pyrite in Carroll and Hot Spring counties; rare earths in Hot Spring and Garland counties; strontium in Pike, Howard, and Sevier counties; and vanadium in Garland County.

PROPORTION OF ACRES PLANTED, COTTON TO CORN AND WHEAT, 1880

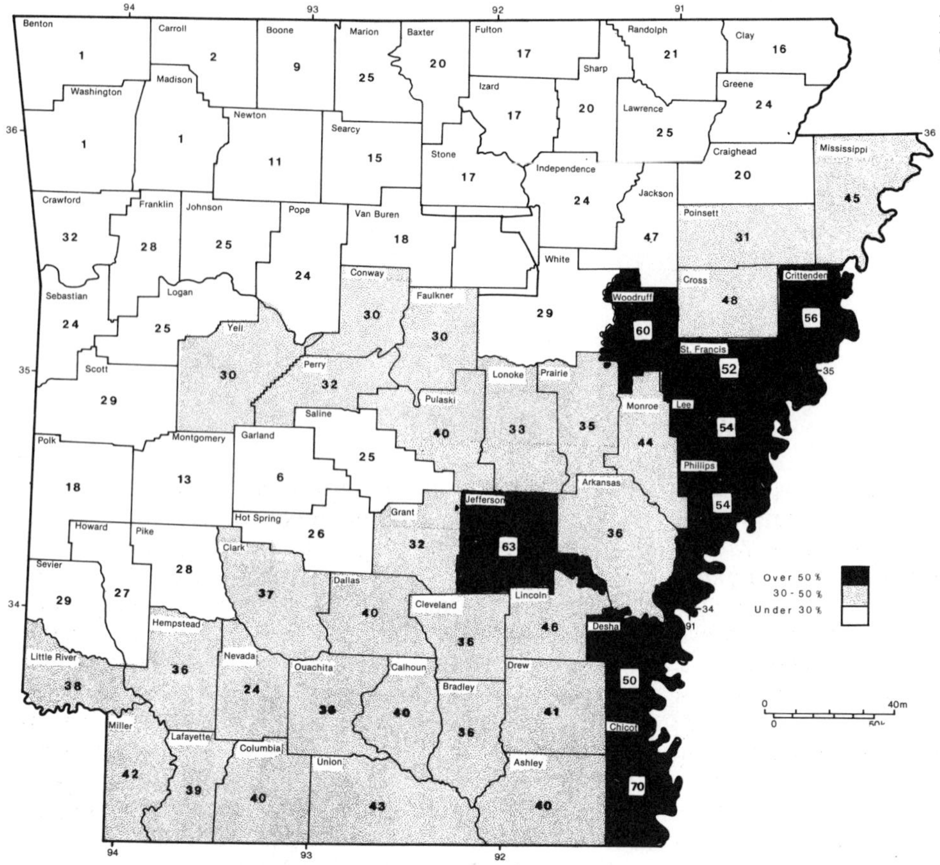

PROPORTION OF ACRES PLANTED, COTTON TO CORN AND WHEAT, 1900

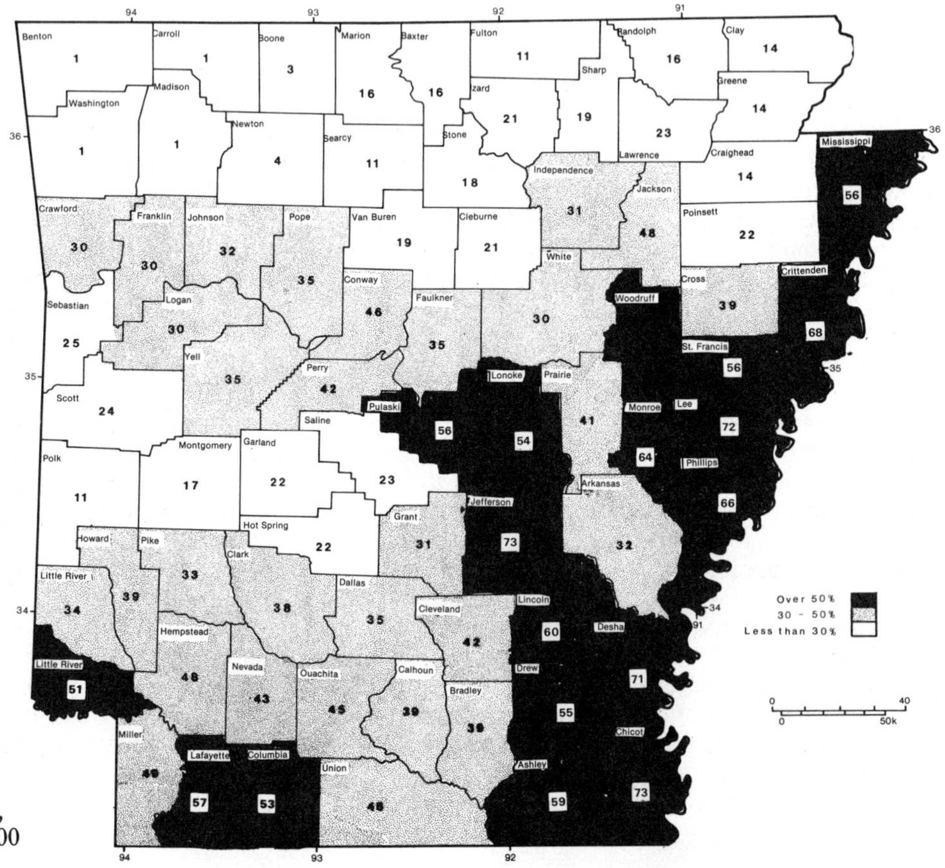

54. NEW SOUTH AGRICULTURAL PATTERNS

54. NEW SOUTH AGRICULTURAL PATTERNS

While the Arkansas economy experienced some diversification after the Civil War, agriculture continued to be its most important component. In terms of production, the Civil War set back farm operations, although within fifteen years after the conflict's end Arkansas crops had surpassed the levels reached at the beginning of the war. Land in farms steadily increased. Between 1870 and 1920, which was the last year the census provided comparable statistics, the number of improved acres being farmed in Arkansas increased from 1,859,821 acres to 9,210,556. Production of row crops remained the central endeavor of farmers, and the antebellum crops of cotton, wheat, and corn occupied most of their attention through the rest of the nineteenth century.

The principal trend in agriculture from the end of the Civil War until 1950 was the spread of cotton culture and its domination of the farm economy. In the 1870's changing conditions began to produce this shift. Railroads made possible greater farm specialization, allowing farmers to purchase cheaper supplies and thus devote more land to cotton, the principal cash crop. The railroads also made it easier to ship products to market from throughout the state and, as a result, encouraged production wherever cotton could be grown. Removal of forests in northeast Arkansas and swampland reclamation along the state's rivers also increased the suitable land available for farming cotton. A comparison of the proportion of farmlands planted in cotton as opposed to corn and wheat in 1880 and 1900 shows the steady shift to cotton. Already in 1880 only six counties devoted less than ten percent of agricultural land to the crop, and these were all in northwest Arkansas, and 30.4 percent of all cultivated lands in the state were planted in cotton. In the delta areas the acreage allocated to cotton was even larger. In Chicot, Jefferson, Crittenden, Phillips, and Lee counties more than half of the farm acreage was devoted to cotton. By 1900 the percent of land devoted to cotton was 32.9. In 1930 cotton land had increased to 52.5 percent. Cotton's share of farmland decreased after 1930, but as late as 1950 still accounted for over one-third of all crop acreage. From 1870 to 1950, Arkansas was one of the chief cotton producers in the nation.

Production of the other traditional crops continued, but their importance declined. Arkansas farmers found competition with cheaper grains imported from the Midwest difficult. In 1880 farmers produced some 24,156,417 bushels of corn. By 1950 the corn grown in Arkansas for grain was negligible. In 1880 wheat farmers, still primarily reflecting the antebellum pattern of production in northwest Arkansas, harvested 1,269,715 bushels of wheat. In 1950, Arkansas produced only 289,454 bushels of wheat.

Cotton, Corn, and Wheat in Arkansas, 1860–1950

Year	Cotton (bales)	Corn (bu)	Wheat (bu)
1860	367,393	17,823,588	199,639
1870	247,968	13,382,145	742,736
1880	608,256	24,156,417	1,269,715
1890	619,494	33,982,318	935,668
1900	704,880	44,144,098	2,449,970
1910	776,879	37,609,544	2,710,000
1920	869,350	34,226,935	2,051,405
1930	1,398,475	27,388,105	153,281
1940	1,351,209	33,762,323	353,443
1950	1,584,307	21,626,026	289,454

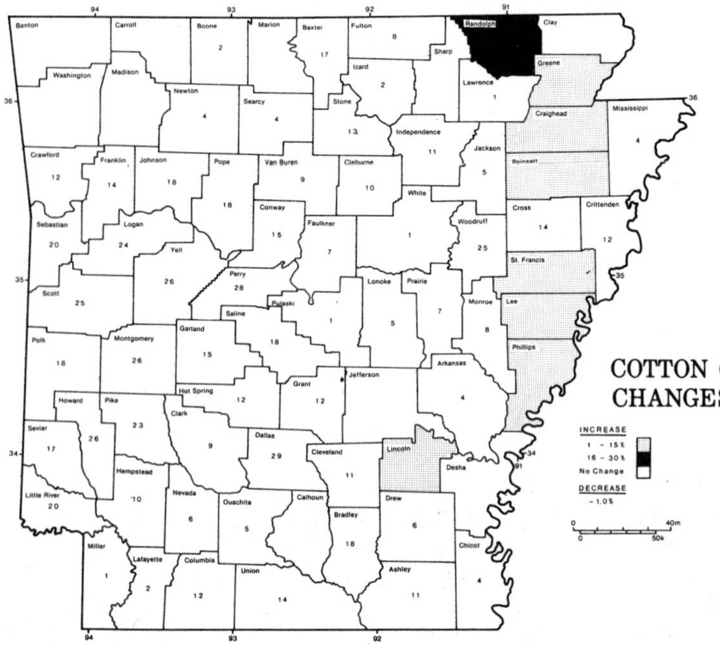

COTTON CROPLAND CHANGES, 1929–1949

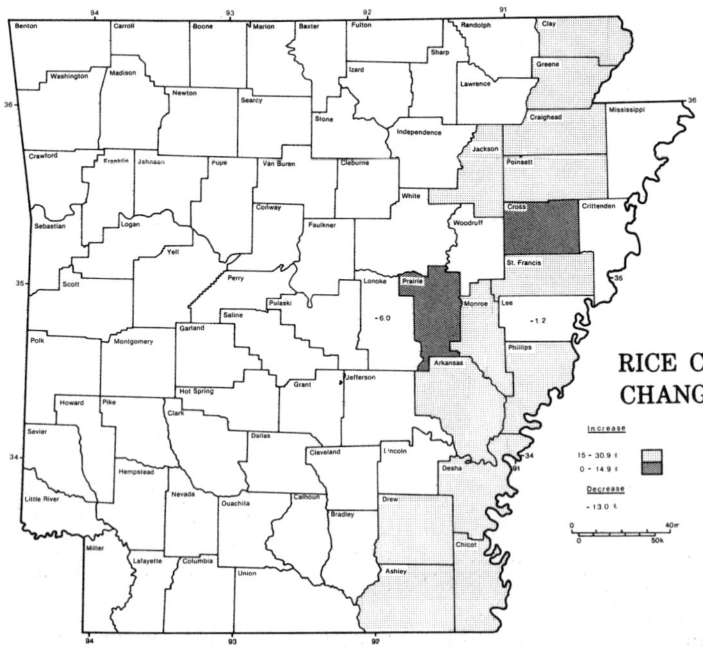

RICE CROPLAND CHANGES, 1929–1949

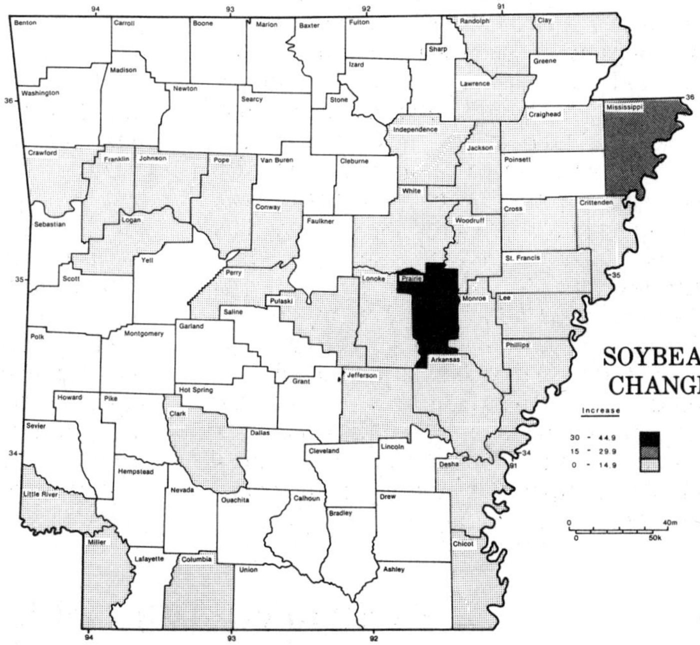

SOYBEAN CROPLAND CHANGES, 1929–1949

55. AGRICULTURAL PATTERNS, 1900–1949

55. AGRICULTURAL PATTERNS, 1900–1949

Decreasing cotton prices between 1870 and 1950 created real problems for Arkansas farmers and forced them to search for alternative cash crops. After 1900 destruction caused by the boll weevil to cotton also pushed farmers to diversify. Up until 1950, however, cotton remained the most important of the state's crops and expanded into many new farming areas, as may be seen in Map 55a.

Indicating the search for alternatives was the emergence at the beginning of the twentieth century of the new crops rice and soybeans. Census figures comparing these crops in 1929 and 1949 show their rise as new and important commodities.

Rice (Map 55b) had been grown in Arkansas as early as 1850, but the crop had never been commercially important. In 1897, after having seen rice under cultivation in Louisiana, William H. Fuller planted three acres of the crop on his farm near Carlisle in Lonoke County. Fuller's success prompted more extensive production. By 1910 farmers had begun to clear forests in an area between the Saint Francis and Cache rivers and the Missouri border and the Arkansas River. East-central Arkansas and the northeast portion of the state became the principal areas involved in the commercial production of rice.

Within a decade, by 1910, Arkansas was the third most important rice producer in the nation, and it has continued in the top three domestic producing states. More critically for Arkansas farmers, the crop offered a viable alternative to cotton. By 1929 it was the fourth crop in value in the state. By 1949 rice was second only to cotton in value. Expansion was rapid as prices remained high until 1920. Lower prices in the 1920's and 1930's, however, slowed down expansion. Higher prices in the 1940's made possible a continued development of the industry.

The second major new cash crop was soybeans (Map 55c). This crop was introduced into the state in the 1920's as a viable alternative to cotton. It required less fertilizer and possessed a strong market as forage and as an oil crop. With no land in cultivation in 1920, by 1930 census reports showed 31,689 acres planted. In 1940 the number had increased to 223,079 acres.

During the 1940's wartime requirements for oil-producing crops led to a rapid expansion of soybean production in Arkansas. To encourage diversion of farmlands to this crop, the goverment raised price supports. With the beans selling for between two dollars and four dollars a bushel, many farmers shifted into producing the new crop. For the landowner, soybeans offered an additional advantage: they could be harvested by machine, thus reducing dependence on human labor. While postwar demand declined somewhat, by 1950 farmers still had diverted additional lands to soybeans, and lands planted in the crop had reached 343,868 acres.

Rice and Soybeans in Arkansas, 1900–1950

Year	Rice (bu)	Soybeans[1] (bu)
1900	8,630[2]	
1910	1,282,830	
1920	6,797,126	
1930	6,958,105	65,869
1940	7,651,231	551,788
1950	19,889,614	6,134,516

[1]Soybeans harvested for beans
[2]Reported in 1900 as lbs.

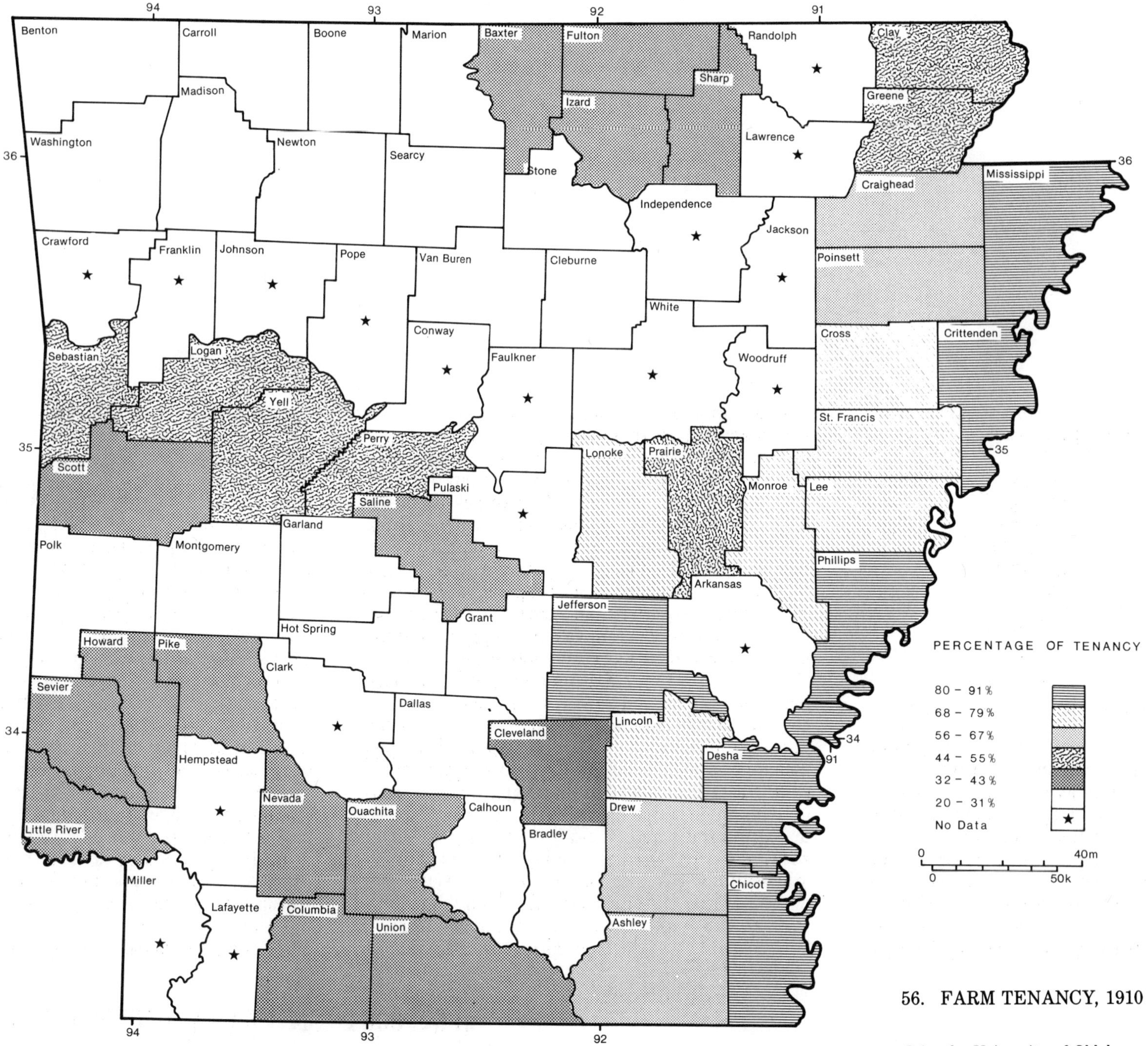

56. FARM TENANCY, 1910

56. FARM TENANCY, 1910

Farming by tenant operators rather than landowners was a significant characteristic of Arkansas agriculture after the American Civil War. In 1860 the state's farms were operated primarily by their owners. After the end of the Civil War in 1865, many owners, especially in the plantation regions of the southeast, subdivided their holdings into smaller units that were farmed by tenants. Such arrangements were typical of the agricultural activities of the freedmen in the postwar years who were prevented by economic conditions and intimidation from acquiring land.

By 1870 several basic types of contracts had emerged. Some farmers simply rented land, paying cash for the use of land and providing their own supplies. A more common practice was to work on shares. This practice had developed during the Reconstruction period as a means of stabilizing labor relations between former slaves and landowners. Sharecropping arrangements varied greatly, but the basic principle was that the tenant provided the labor while the landowner provided land, implements, draft animals, and supplies. The two then shared the crop, with contracts that varied widely. The tenant who was able to furnish everything but the land received as much as three-fourths of the cotton and one-third of the corn and other crops. Tenants who furnished only labor received as little as one-third of the crop. Some distinction has been made between share-tenants and sharecroppers, but the only real differences among those working on shares was the proportion of the crop that they could negotiate. There was also some wage labor, although this practice was not greatly favored.

By 1880, a decade later, 31 percent of all farm operations in Arkansas were carried on by tenants. In the delta regions that had been dominated by antebellum cotton plantations the proportion was as high as 90 percent. Economic conditions during the rest of the nineteenth century worked to undermine smaller agricultural operators and force them into tenancy, thus adding large numbers of whites to what had been primarily a black tenant force. In 1890, 32 percent of all farms were tenant-operated. By 1900 the number was 45 percent. The 1910 census provided the first careful analysis of the problem and showed that for the first time as many farms were operated by tenants as owners. The rise of tenancy continued until it reached a high point in 1930 when 63 percent of farms were tenant-operated.

The trend towards increasing tenancy has been reversed since 1930, primarily because of the introduction of agricultural mechanization that has made possible farm consolidation and the dispersal of the tenants. Tenantry has closely paralleled farm size in the state. In 1900 the average farm size in Arkansas had been 126 acres, but by 1930 that figure had dropped to 66 acres. The introduction of mechanization and the impact of the Great Depression resulted in an increase in the average size of farms to 265 acres by 1978. With this shift the number of tenants steadily declined. By 1940 the proportion of tenant farmers had dropped to 53 percent, by 1950 to 38 percent. More significant changes took place after 1950. By 1969 the number of tenant farms was only 13 percent of the total, and by 1978 it was 11 percent.

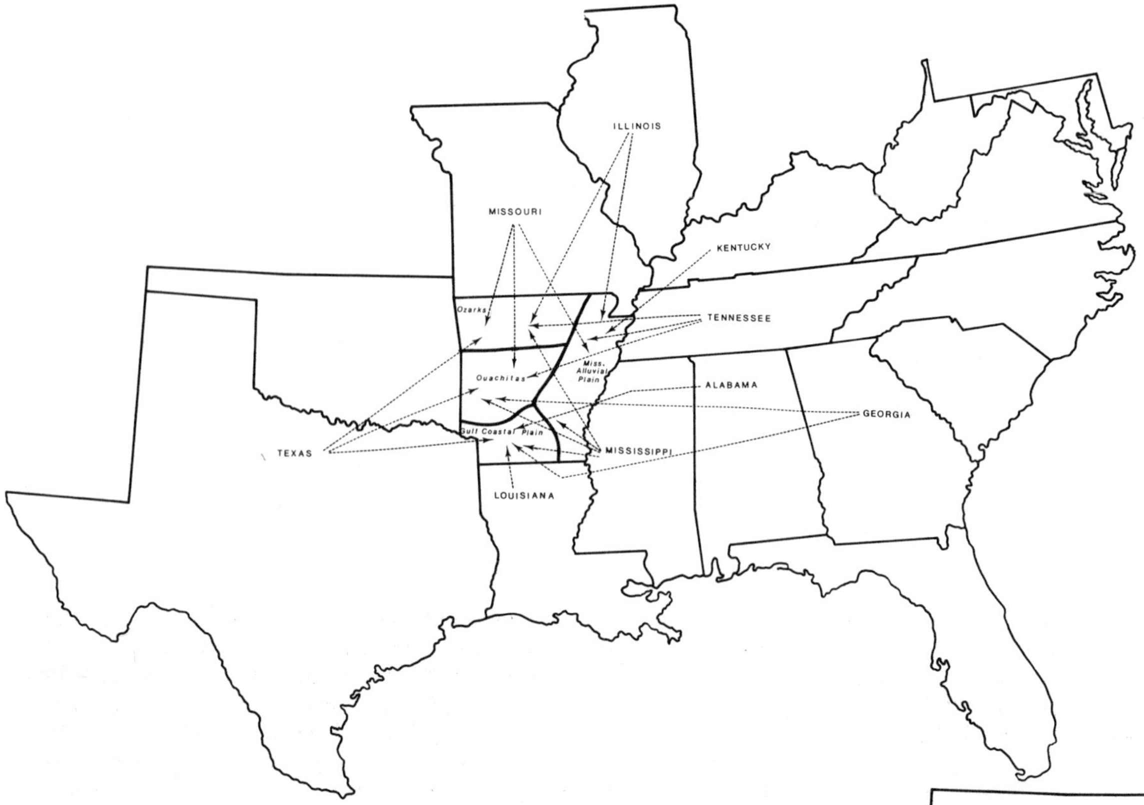

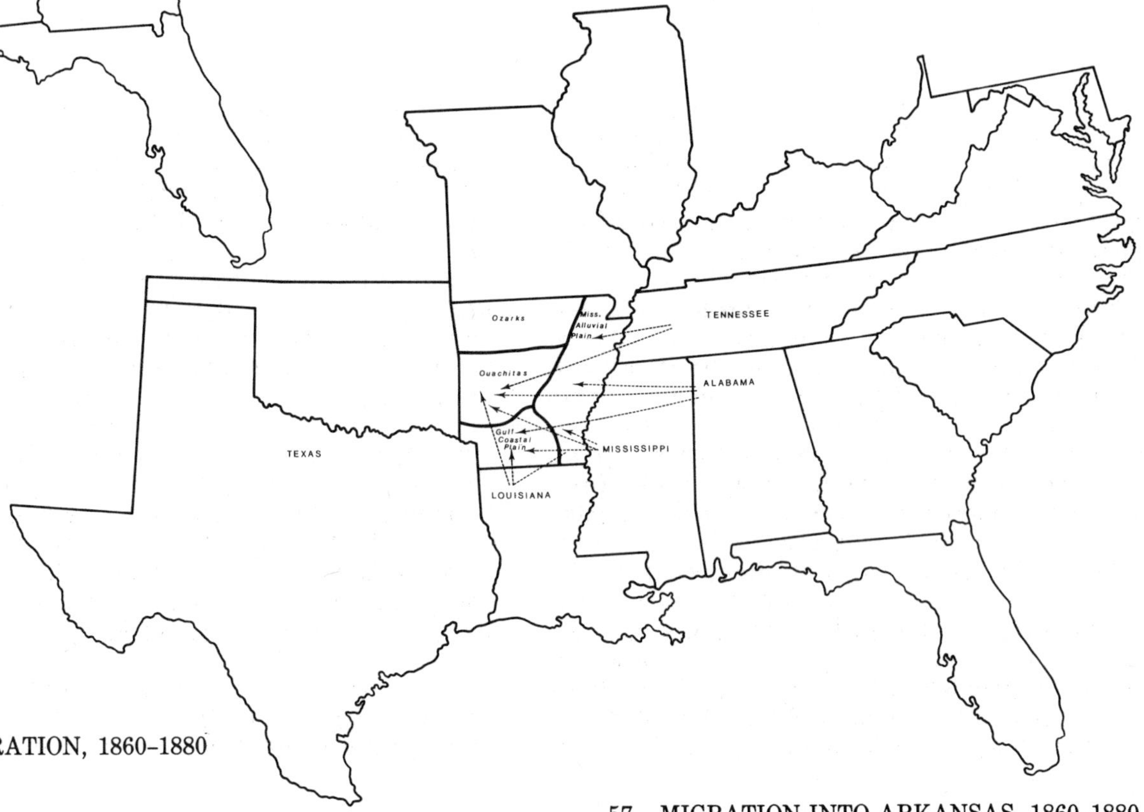

57. MIGRATION INTO ARKANSAS, 1860–1880

57. MIGRATION INTO ARKANSAS, 1860–1880

Economic opportunities in Arkansas brought continued immigration to the state during the period from 1860 to 1880, although the basic trends established during the antebellum years were modified. As might be expected, the volume of immigrants of both races declined during the Civil War years, but increased again with peace.

Between 1860 and 1870, 87 percent of white immigrants came from the old slave states. Of the free states, only Illinois contributed a significant number to the state's population. There was also a shift in the relative importance of the old slave states as contributors to Arkansas's population. In this decade Missouri became the most important source of immigrants, followed by Texas, Tennessee, and Mississippi. The new importance of Missouri and Texas may have been the result of wartime migrations out of the state: perhaps as many as 20 percent of the people who moved to Arkansas from Missouri and Texas in the 1860's had been in the state before the war. The movement out of Mississippi represented a general trend among white small farmers, who left the older states of the Deep South searching for new opportunities and to escape from what they feared would be a violent racial situation.

In the period from 1870 to 1880 the state experienced its largest volume of immigration. The domination of that movement by Missouri, Tennessee, and Texas continued. In this period as many as 25 percent of the immigrants were former residents returning to Arkansas. It is more difficult to explain the return of people to Arkansas during this period than the movement in the 1860's. Possibly the end of Reconstruction in 1874 paved the way for the return of more wartime refugees. In addition, this movement may reflect dissatisfaction with the new farming frontiers on the plains of Texas that had been opened in the postwar years.

In the 1860's black migration came chiefly from Texas, Tennessee, Mississippi, and Alabama, in that order. About 80 percent of this movement took place after the end of the war in 1865, and about one-half of the immigrants had lived in Arkansas previously. As in the case of the white population, this trend probably reflected the return of wartime refugees to their homes.

Between 1870 and 1880 the volume of black immigration increased, and the sources shifted. Tennessee became the most important source of new Arkansans, followed by Mississippi, Louisiana, and Alabama. While this movement probably reflects a continued return of wartime refugees, at least in part, it also indicates the increased mobility experienced by the postwar freedmen. New lands in Arkansas, which it was rumored promised three bales of cotton per hand, may have attracted the former slaves who were seeking economic opportunities. Central Arkansas and the plantation regions on the south and east became the places where these immigrants concentrated. The agricultural emphasis of all of these counties except Pulaski further implies a movement in search of economic options.

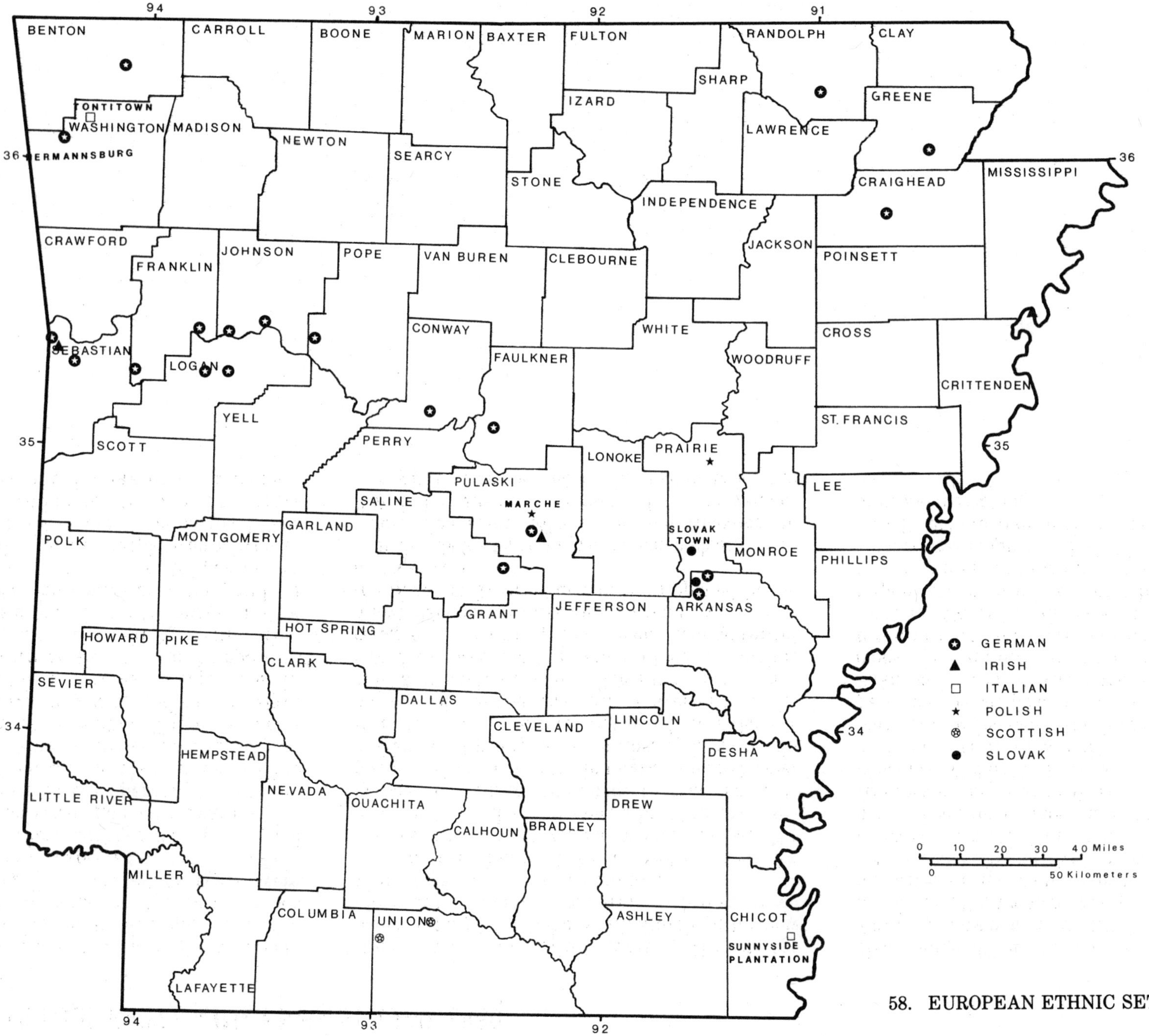

58. EUROPEAN ETHNIC SETTLEMENTS

58. EUROPEAN ETHNIC SETTLEMENTS

Arkansas has never attracted a large number of foreign immigrants. Nonetheless, there have been important foreign communities established in the state, and their presence has added a more cosmopolitan flavor to what would otherwise have been ethnically homogeneous black and white farming populations.

The European population of Arkansas peaked in 1900, when there were 14,289 foreign-born people living in the state, about one percent of the total population. Thoughout the state's history the nature of the immigrant population has been relatively stable. Except in 1890 the majority of foreign Arkansans came from Great Britain or the British Isles. This percentage was followed by German immigrants, who led the list in 1890. In 1900 those two nations were followed, in descending order, by British America (Canada, for the most part), Switzerland, Italy, Austria, France, Sweden, Bohemia, Russia, and Poland. A scattering of additional immigrants came from every continent of the world.

European immigration to Arkansas had taken place from the beginning of statehood. In 1836 sixty Rhinelanders, led by the Reverand Gustavus Klingelhoffer, settled near Little Rock. In the 1850's a German settlement, known as Hermannsburg, was established in Washington County. The total number of European immigrants, however, was small.

The majority of the European immigrants who arrived in Arkansas came during the last three decades of the nineteenth century. Many were Germans, Austrians, or Swiss who came as a result of efforts by the Arkansas railroads to attract immigration, especially from among German Catholics. The railroads had received large federal land grants, and hoping to sell their lands, the roads sent agents to Europe to recruit settlers. The result of these efforts was the establishment of several German-speaking settlements in the upper Arkansas River valley and on the Grand Prairie. By the turn of the century one-half of the foreign population in Arkansas was in these Arkansas River communities.

Other foreign elements came in at this time as well. Slovak Town, in Prairie County, was settled by Slovaks, Bohemians, and Russians recruited in the American Midwest by the Slovak Colonization Company, which owned ten sections of land on the Grand Prairie. In 1897 an Italian community was established in Washington County at Tontitown. The settlers were members of eighteen families from northern Italy who were led to the United States by Father P. Bandini. This group of 562 immigrants had left Genoa in 1895 and landed in New Orleans. They settled first at Sunnyside Plantation in Chicot County, where they worked as contract laborers or indentured servants. When they had fulfilled their contracts, Bandini led them to Washington County, an area more like their homeland. The Italians at Tontitown were joined later by other European settlers. Their community engaged primarily in fruit production, and they particularly excelled in growing grapes. Polish communities were located at Marche in Pulaski county and in Polania, now present-day Des Arc, in Prairie County.

BLACK POPULATION, 1860

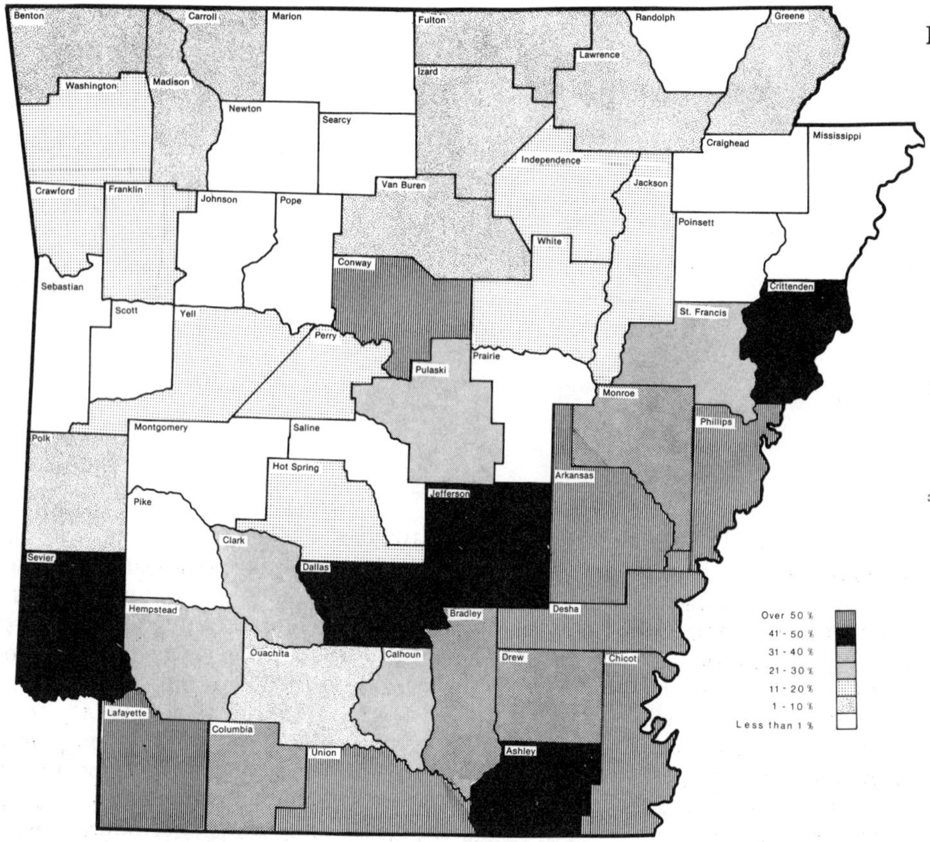

BLACK POPULATION, 1900

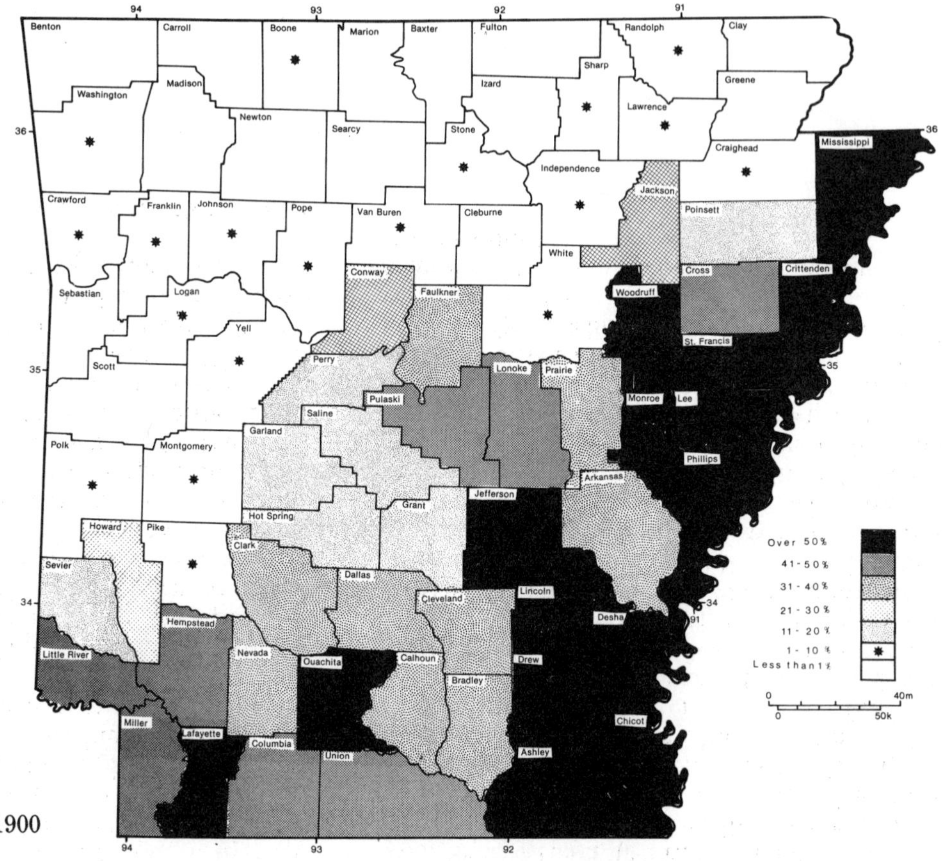

59. BLACK POPULATION
SINCE THE CIVIL WAR

59. BLACK POPULATION SINCE THE CIVIL WAR

Blacks have always constituted the largest minority population in Arkansas. In 1860 there were 111,115 black inhabitants, roughly 25 percent of the state's population of 435,450. Most of these people were slaves. Consequently, they were tied to the counties with a strong plantation economy. The majority were rural: in 1860 there were only 3,273 blacks in the state's seven cities—about 3 percent of the state's black population. Six counties had black majorities: Arkansas, Desha, Union, Chicot, Phillips, and Lafayette. Union and Lafayette were part of the plantation region along the Red and Ouachita rivers. The other four counties lay along the Mississippi. The four counties with the largest black populations were Phillips County, with 8,945; Chicot County, with 7,512; Jefferson County, with 7,158; and Union County, with 6,331.

By 1900 the black population had changed appreciably. Of the state's total population of 1,311,564, 366,856 were black. This represented an increase to almost 28 percent of the total population and demonstrated that Arkansas was the home of a fairly stable black community by the turn of the century, a community that not only kept its population but also attracted immigration from other states. During the four decades following the Civil War the state's black population increased by 57.5 percent per decade, while the white population's rate of increase was 47.8 percent.

The great majority of black Arkansans remained in rural areas and on the farms, but as was true of the white population, a trend had emerged towards urbanization. By 1900, 37,171 blacks lived in urban areas, a little over 10 percent of the total black population of the state.

Since 1900 the rapid growth of the black community has been reversed. In 1980 the black population was only 373,025, a little over 6,000 more blacks than were in the state in 1900. While the increases in the black population have been less than 1 percent per decade, the white population in the same period grew an average of 12.5 percent per decade. A major trend among Arkansas's black population has been migration into the cities. In 1980 the state's blacks were 65.8 percent urbanized, compared with only 48.6 percent of the white population.

Arkansas's Black Population Since the Civil War

Year	Total	Percent Increase/ Decrease	Percent of Population
1860	111,259		25.6
1870	122,169	9.8	25.2
1880	210,666	72.4	26.3
1890	309,117	46.7	27.4
1900	366,856	18.7	28.0
1910	442,891	20.7	28.1
1920	472,220	6.6	27.0
1930	478,463	1.3	25.8
1940	482,578	0.9	24.8
1950	426,639	-11.6	22.3
1960	388,787	-8.9	19.9
1970	352,445	-9.4	18.3
1980	374,000	6.1	16.3

1870-1900

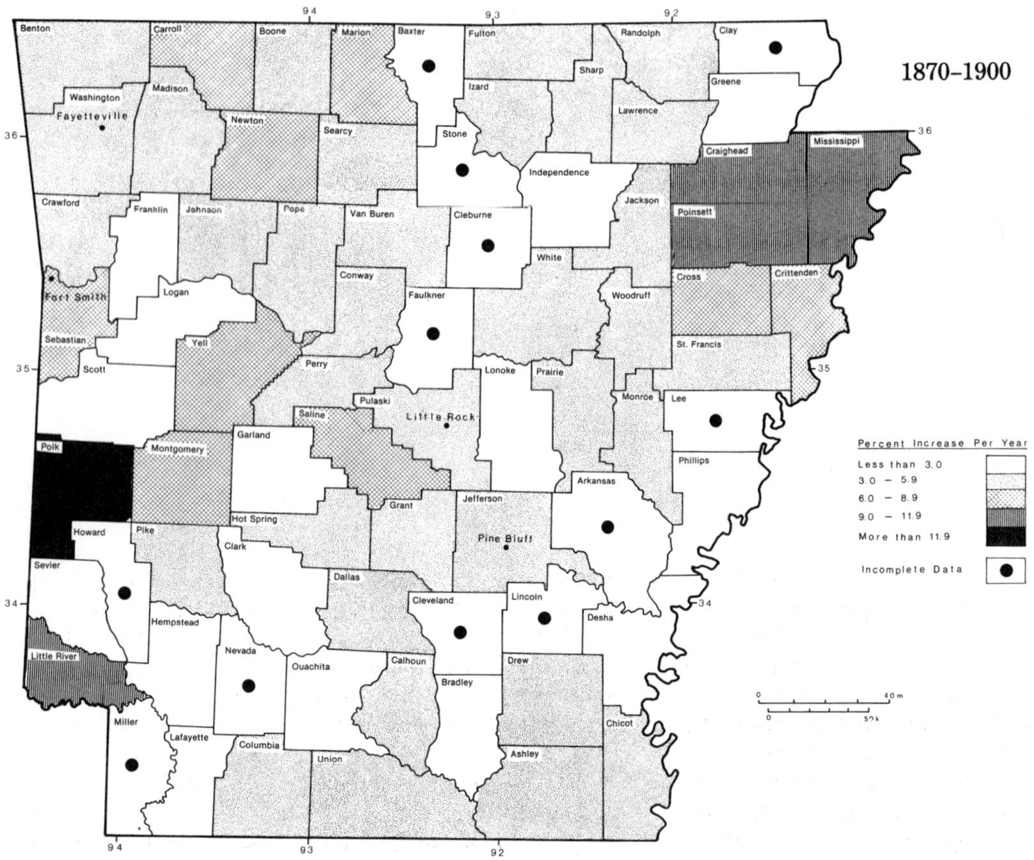

1900-1940

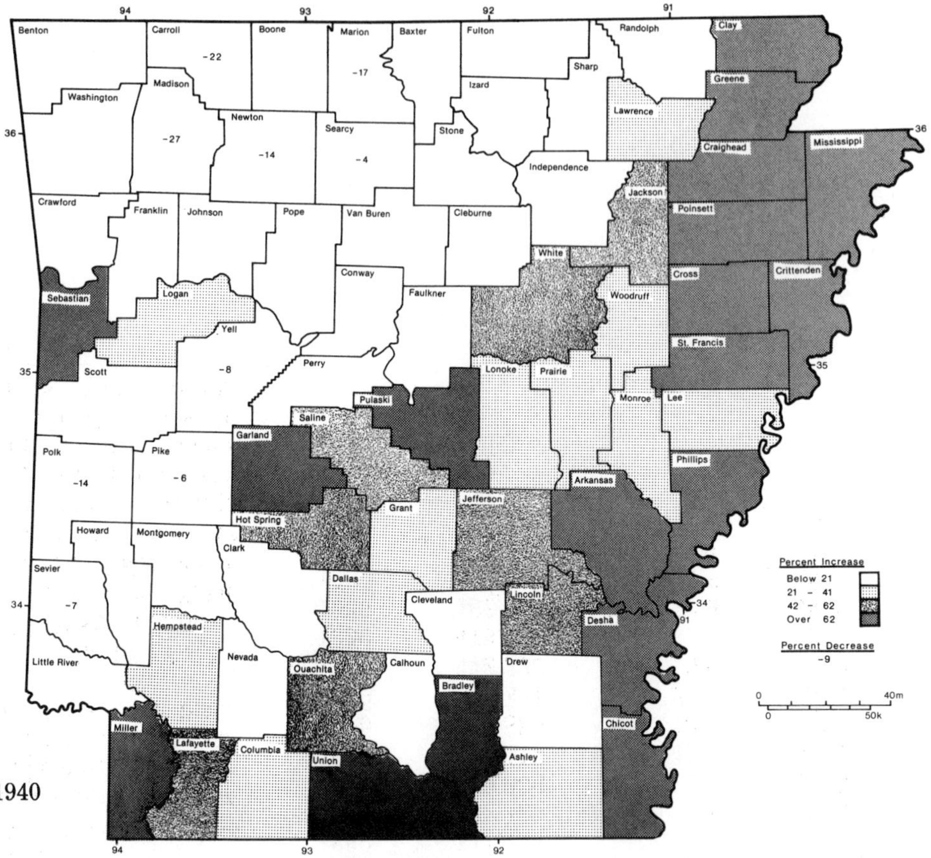

60. GENERAL POPULATION TRENDS, 1860-1950

60. GENERAL POPULATION TRENDS, 1860–1950

Between 1860 and 1950 the general population in Arkansas grew, although this basic trend was reversed in the decade of the 1940's. The growth rate appears to have been connected with general economic trends in the state, with the plight of agriculture being a major force determining population.

From the end of the Civil War until 1890, Arkansas's population expanded rapidly, reflecting the opportunities presented by the general expansion of the state's economy. New farms and farmlands, railroad construction, and the emergence of the timber industry proved attractive to outsiders, and the migration of new inhabitants from the older southern states and foreign immigration helped to bolster the population expansion that was general throughout the state. The farm population kept pace with the urban population, despite a slightly faster rate of growth for the latter. By 1890 the population had more than doubled since 1860 and had passed 1 million.

Agricultural problems in the 1890's appear to have stifled the rapid population expansion. Low growth rates during the decade and actual declines of population in some rural areas suggests the link between farm problems and slower growth. Slow growth in the 1920's and 1930's also suggest a parallel with the poor economic situation in the state. Rural growth rates continually fell behind that of the general population. The worst decade was the 1920's, when rural population increased at a rate of only 0.7 percent, compared with a rate throughout the state of 5.8 percent.

By the 1940's population had begun actually to decline. Changes in the structure of agriculture, which remained the state's most important economic activity, pushed many individuals off the farms. In particular, the shift from cotton planting to soybeans, which could be mechanically harvested, began to deprive many in the rural population of any economic opportunities. Also, war industries located outside of the state pulled Arkansans into other regions. While the state at large suffered a population decline of 2.0 percent, the rural areas suffered a population reversal of 14.8 percent.

The basic population trends between 1870 and 1950 showed the importance of the agricultural economy to Arkansas. They further showed the problems within that sector of the economy. Dependent on the farm and farm products, Arkansas declined because of changes in the nature of farming and farm life.

General Population Trends, 1870–1950

Year	Total	Percent Increase or Decrease
1870	484,471	
1880	802,525	65.6
1890	1,128,179	40.6
1900	1,311,564	16.3
1910	1,574,449	20.0
1920	1,752,204	11.3
1930	1,854,482	5.8
1940	1,949,387	5.1
1950	1,909,511	−2.0

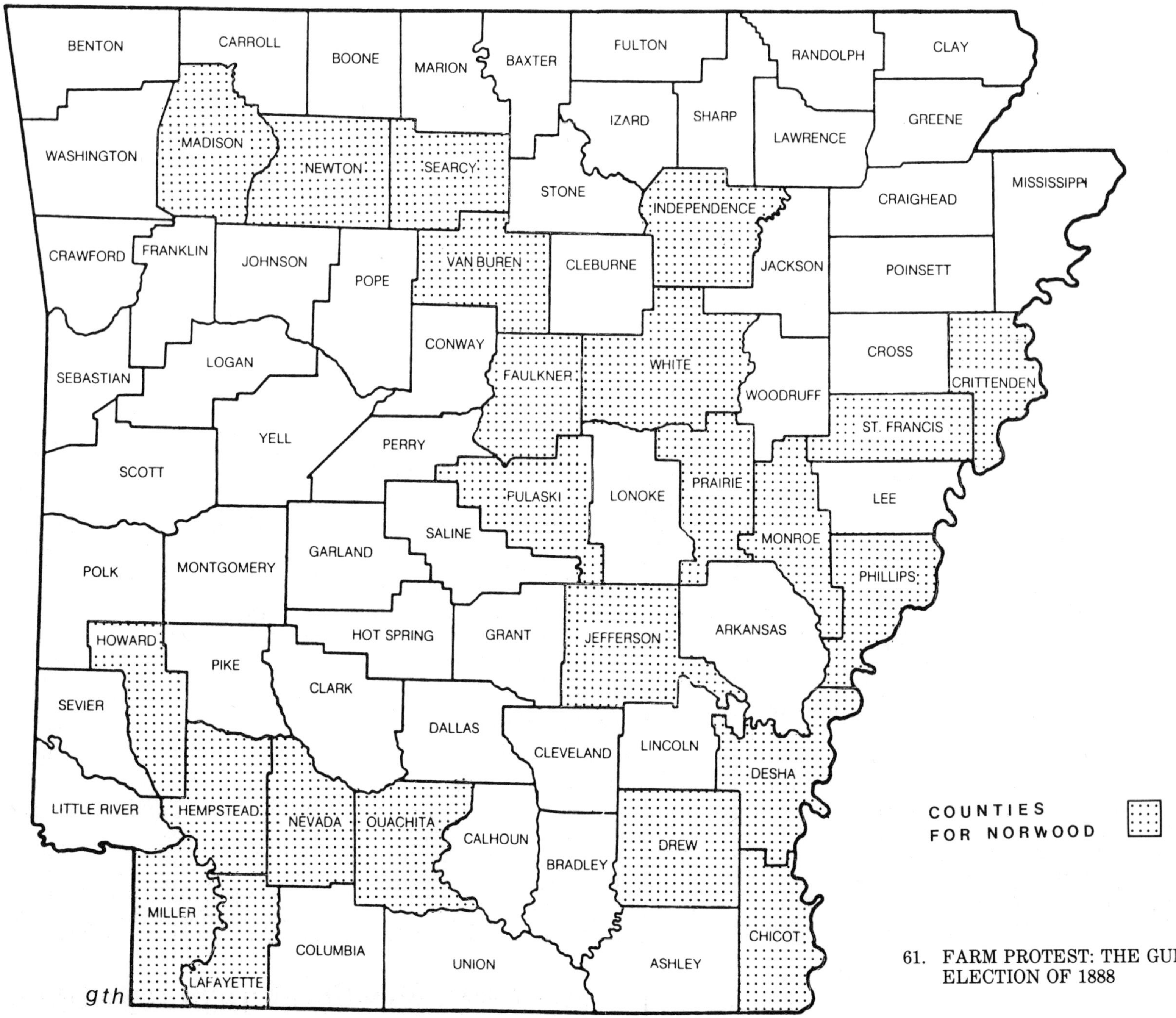

61. FARM PROTEST: THE GUBERNATORIAL ELECTION OF 1888

61. FARM PROTEST: THE GUBERNATORIAL ELECTION OF 1888

A major force in Arkansas politics after the Civil War was the pernicious agricultural depression. Declining commodity prices hurt all farmers, rich and poor. The poor, however, tended increasingly to demand some sort of political remedy for their problems. In the 1870's many had joined the Farmer's Alliance, the Grange, the Brothers of Freedom, the Greenback party, or other such groups. In 1880 the Greenback party challenged the Democratic party seriously in Arkansas and obtained a fourth of the vote for their gubernatorial candidate.

Farm protest became more organized and presented a greater political challenge after the organization of the Agricultural Wheel near Des Arc in Prairie County in 1882. The National Agricultural Wheel was organized in the state four years later, at Litchfield on July 28, 1886. The Wheel made nominations for the 1886 elections, but did poorly. In 1888 many Wheel leaders helped to organize the Union Labor party, which combined with the state Republicans and various farm organizations. In 1888 the Union Labor party nominated Dr. Charles M. Norwood of Prescott. The party was a remarkable combination of voters in nineteenth-century Arkansas, with blacks from the plantation counties, who were traditionally Republican, alongside small farmers in counties with primarily white populations.

The 1888 election was marked by considerable fraud and violence. Union Labor candidates claimed victory, but in the official tabulation of the vote the Democratic candidate, James P. Eagle, was declared the winner. Eagle's vote was 99,214 to Norwood's 84,213. Norwood carried twenty-three counties, obtaining support in plantation counties in the southwest and southeast and in several northern counties.

The 1888 election was a turning point in Arkansas politics. Because of the threat presented by Union Labor's biracial coalition, the Democratic leadership responded with actions that broke it up. Violence was used to intimidate Union Labor voters. Party leaders also initiated legal means to disfranchise blacks and thus destroy a major part of the new party's strength. The election law of 1891 and the poll-tax amendment to the state constitution adopted in 1892 were major components of the attack on black voters. The election law replaced printed ballots on which the party was identified by the color of the ballot with the Australian, or secret, ballot. The poll tax required prepayment of the tax and elaborate record keeping to be able to participate in elections. After 1892 black participation in elections declined. By the first decade of the twentieth century the Democratic party's adoption of the white primary truly disfranchised black voters by eliminating them from voting in the only meaningful election in the state. In the face of these attacks many whites who had cooperated with blacks in this party now turned away. The 1888 election would be the last major threat to Democratic hegemony in Arkansas until the 1960's.

61. FARM PROTEST: THE GUBERNATORIAL ELECTION OF 1888

PART EIGHT
MODERN ARKANSAS, 1950 TO THE PRESENT

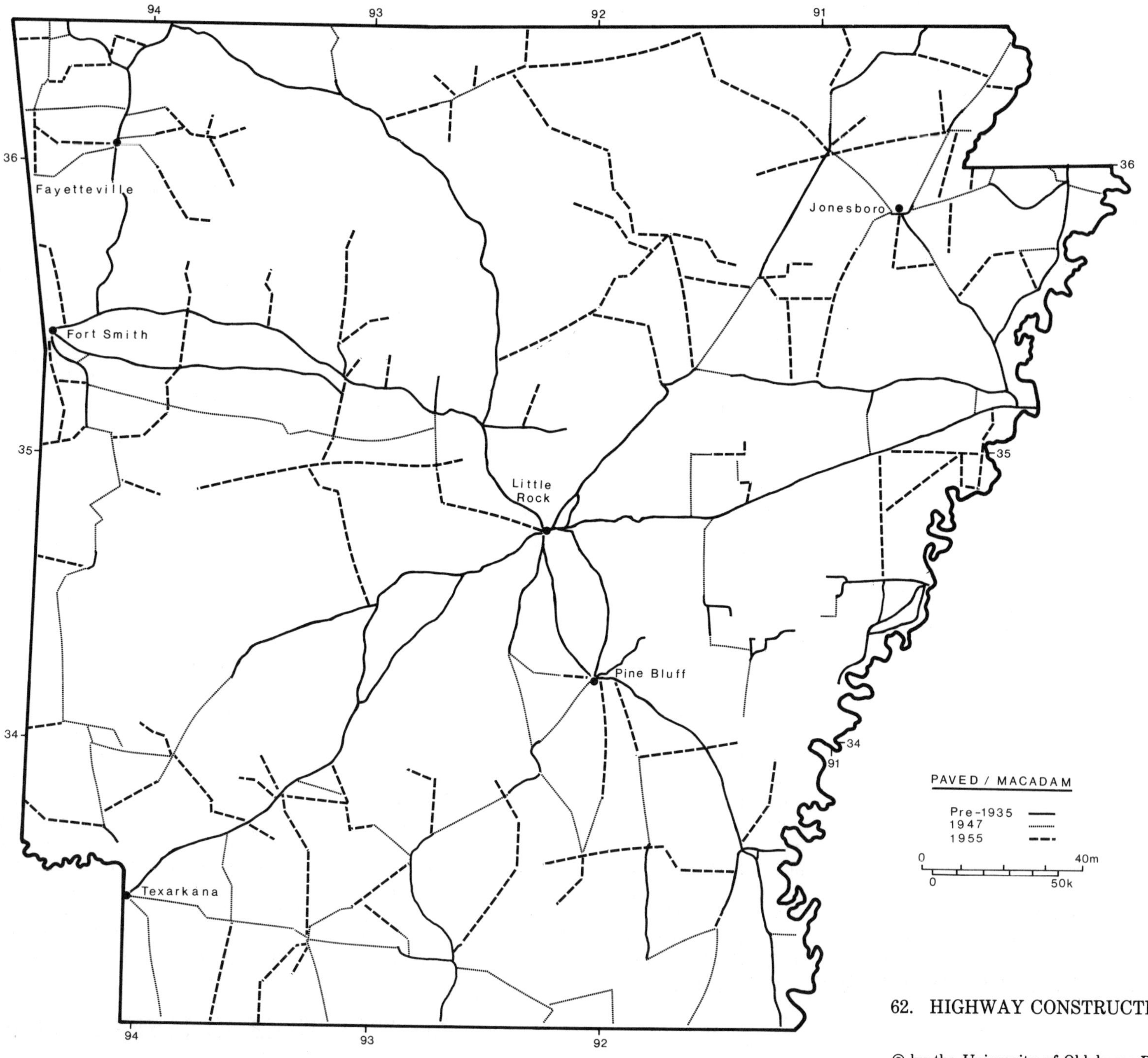

62. HIGHWAY CONSTRUCTION, 1925–1955

62. HIGHWAY CONSTRUCTION, 1925–1955

A major new solution to the problems of transportation that plagued nineteenth-century Arkansas arrived with the automobile in the early twentieth century. More effectively than steamboats or railroads, the automobile almost from the beginning promised to integrate isolated rural Arkansans into their regional and national markets and communities. Although construction began early in the century, the full impact of the new technology would not be realized until the 1950's, and, in some ways, the automobiles and trucks made possible an economic revolution at that time.

The first state law providing for the construction of highways was passed in 1907. It authorized local road-improvement districts that could fund construction through bonds and taxes. The district system never achieved good results, however, and as a consequence, in the 1920's the state created a new system in which the state supervised the roads.

In 1923 the Harrelson Road Act used revenues from a gasoline and motor-oil tax and from auto licenses to fund a state system. In the 1925–1926 biennium the new highway department reported that 8,345.5 miles of roads were under state maintenance. Of this only 734 miles were surfaced, about 9 percent of the total. In 1927 the state took over the existing improvement districts and assumed and funded their debts. The state then began a four-year construction program that was funded by issuing bonds. By 1931 the bonded indebtedness for the program had reached $90 million. This first great expansion ended with the depression. The expenditures did not greatly add to the total miles under state care, but they significantly increased the roads that were paved. In 1935–1936 the state maintained 8,927 miles of roads, of which 2,179 were paved.

There was little expansion of the road system during World War II. By 1948 there were 9,742 miles of road in the state, of which, 4,283, or 44 percent, were paved. After the war another great effort at building highways began. Voters approved the issuance of $28 million in highway construction bonds in 1948. These bonds made possible a four-year construction program begun in 1949. In the first three years three-fourths of the amount of authorized bonds were issued. Because of a scandal connected with the bonds, however, the fourth installment of bonds was cancelled. A major result of this scandal was the passage of Amendment 42 to the state constitution, providing for an independant highway commission. The amendment's provision for a commission of five members serving staggered ten-year terms was to ensure that no governor would ever appoint more than two members in the usual two terms, or four years, in office. Orval Faubus, however, who served twelve years between 1954 and 1966, dominated the commission. The state's road system showed some progress under the postwar program and in the first year of the new commission. By 1955 the state had 10,038 miles of road, of which 6,907, 69 percent, were paved.

The accompanying map shows the highway system as it existed in 1955 and represents the basic pattern that continues to exist. The number of miles continue to expand, however, and a real revolution in road construction took place in Arkansas after 1956, when the U.S. Congress provided for the construction of the defense highways usually known as the interstate highway system. Because the federal government provided 90 percent of the costs of highways within this interstate system, state revenues were freed for work on Arkansas's secondary roads. By 1982 the state highway commission reported maintenance of some 16,095 miles of road, of which 15,332, or 95 percent, were paved.

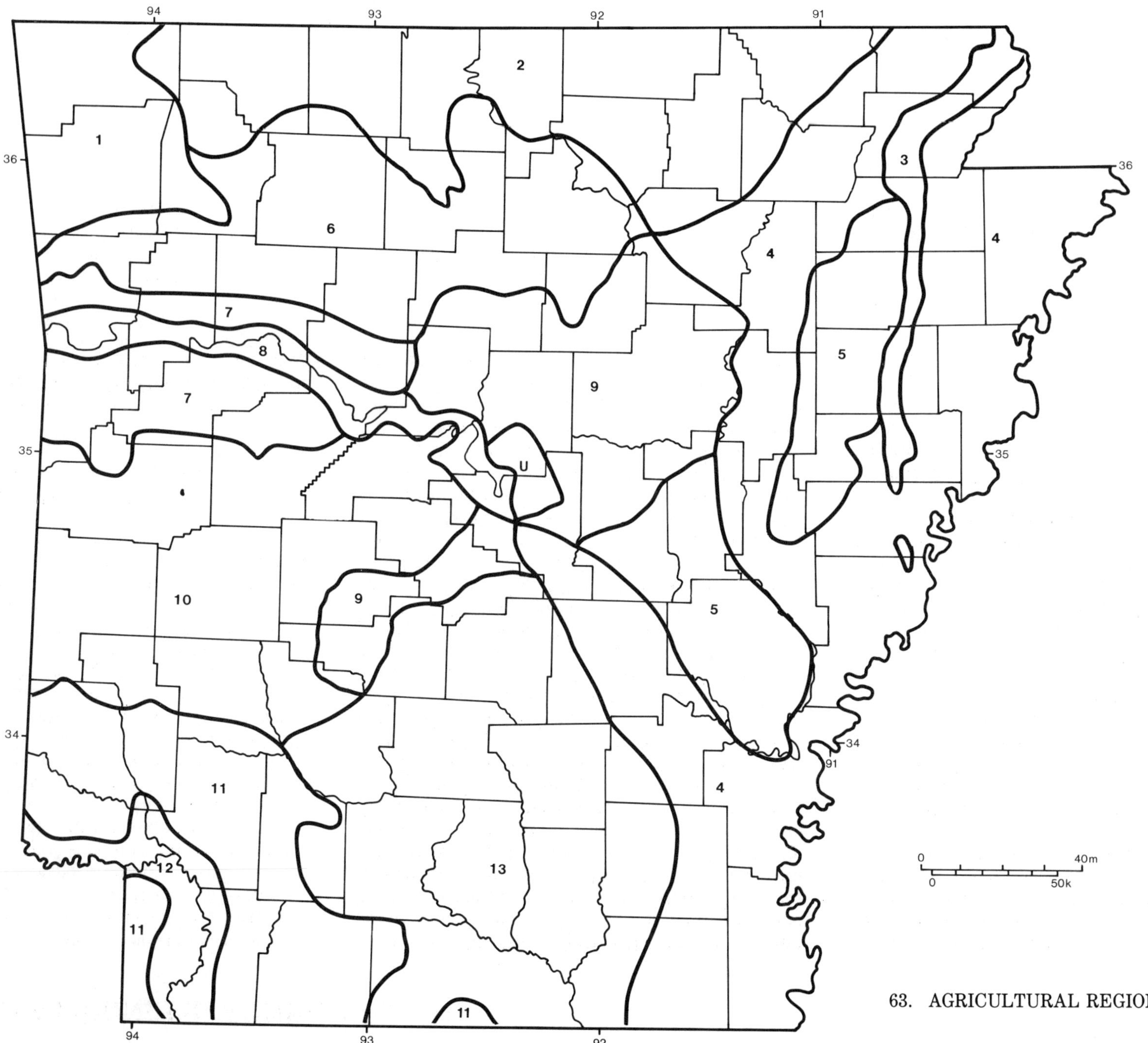

63. AGRICULTURAL REGIONS

63. AGRICULTURAL REGIONS

Agriculture has continued to be the principal industry of Arkansas in the twentieth century. In 1980 its products constituted approximately two-fifths of the value of all goods produced in the state. The basic structure of agriculture has changed significantly, however, as a result of changing markets and new markets made accessible by highway development. The value of livestock produced for sale is now more important relative to traditional row crops. In 1940, the first year the census provided figures on crop and livestock values, Arkansas farm crops were valued at $142,087,620, while the sales of livestock and livestock products were valued at $21,270,727. Beginning in the 1940's, the development of the state's poultry industry spurred a rapid increase in the value of livestock and poultry products. As late as the 1974 census report the economic value of crops was still greater—$1,195,945,000 over $847,135,000—but by 1984 for the first time the value of animals and of animal products had increased beyond that of crops—$1,469,879,000 to $1,356,618,000.

Basic row crops have continued to be important, however. In 1980 soybeans were the state's most valuable crop. Farms grew 31,500,000 bushels of this commodity. In 1980, Arkansas was the nation's leading rice producer, with 52,615,000 hundredweight harvested in 1980. Cotton is the third crop in order of value among Arkansas crops. The state grows approximately 7 percent of the nation's total. In 1980 this was 444,000 bales.

In 1955 the University of Arkansas School of Agriculture identified thirteen major agricultural regions in the state. While the importance of specific products has changed, the basic agricultural activities in each region have not. These regions illustrate the diversity of agricultural activities and potential in Arkansas:

Region 1: the Springfield Plateau. There is more cropland than woodland in this region, where farms specialize in poultry, dairy cattle, and beef cattle. In addition, farmers grow fruits, such as grapes and apples, and vegetables, primarily tomatoes and greenbeans.

Region 2: the Salem and eastern Springfield plateaus. The dominant farm interests here are beef cattle and dairies. Pasture dominates the western part of the region, where poultry production is also significant. Vegetables and strawberries are raised in the center of the region and cotton on the east.

Region 3: Crowley's Ridge. This region is the home of general farming. While there is still some cotton and fruit production, beef cattle and dairying are the predominant interests.

Region 4: the Mississippi Alluvial Plain. This area is the home of the state's major cash crops: cotton, soybeans, wheat, corn (in the north), and rice. There is some livestock raising, but it is less significant than in other regions of the state.

Region 5: the Mississippi Alluvial Plain. This region is distinctive because of its emphasis on rice production, which is made possible by the existence of irrigation water and a compact subsoil. Rice growing is supplemented by some soybean, cotton, and beef-cattle production.

Region 6: the Boston Mountains. This general farming region is mostly woodland rather than cropland. Livestock raising is the chief endeavor. Beef cattle, hogs, and chickens are the most common commercial products.

Region 7: the Western Arkansas Valley. This old agricultural region formerly emphasized cotton production, but shows greater diversification today. It produces beef and dairy cattle, corn, vegetables, and fruit.

Region 8: the Upper Arkansas Alluvial Region. Soybeans, corn, oats, and grains are the chief crops of this diversified farming area where the more-productive farms are on the alluvial bottomlands.

Region 9: the Central Agricultural Region. Dairying, beef-cattle and poultry production, and cotton, vegetable, and fruit farming dominate this area.

Region 10: the Ouachita Mountain region. Farming in this area is limited to pasturage, hay, corn, and some livestock production. Most of the acreage is in woodland.

Region 11: Southwestern Gulf Coastal Plain. This region sustains cotton, corn, and pasture production. Vegetables and fruit are also important. There are significant beef-cattle, dairy, and poultry operations as well.

Region 12: Bottomlands of the Red and Little rivers. This region is dominated by cotton and beef cattle.

Region 13: Central Southern Gulf Coastal Plain. This portion of the plain is more heavily forested than region 11. Agricultural products include beef cattle and cotton, with some vegetables.

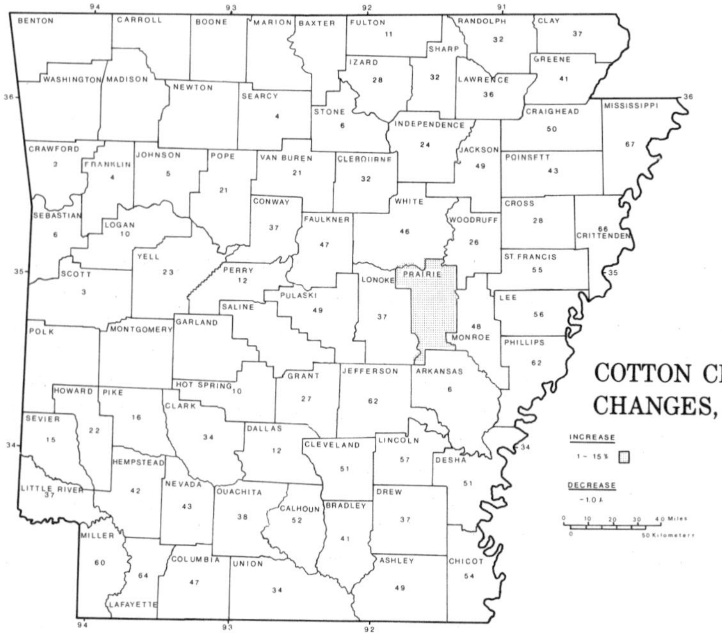

COTTON CROPLAND CHANGES, 1949–1969

INCREASE
1 - 15%
DECREASE
-1.0%
0 10 20 30 40 Miles
0 50 Kilometers

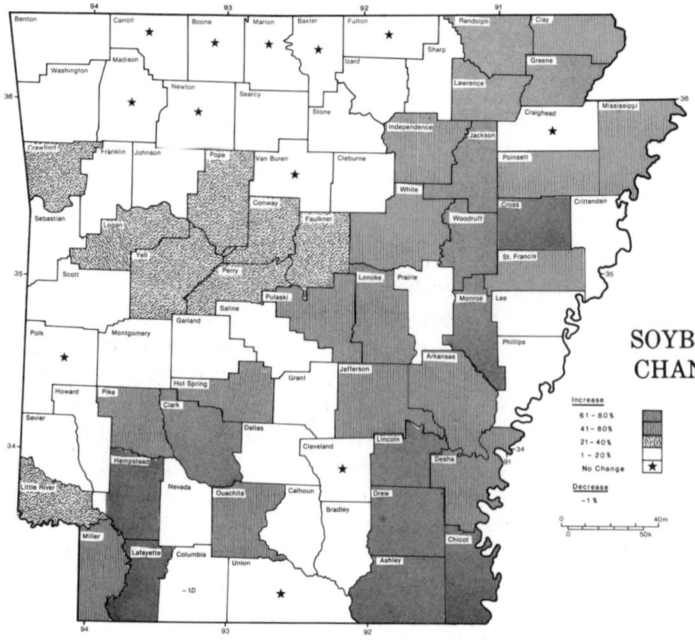

SOYBEAN CROPLAND CHANGES, 1949–1969

Increase
61 - 80%
41 - 60%
21 - 40%
1 - 20%
No Change
Decrease
-1%
0 40m
0 50k

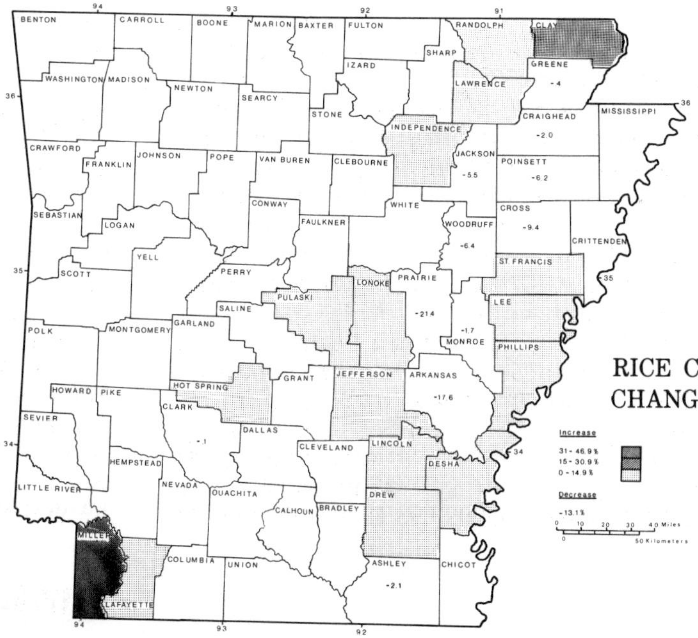

RICE CROPLAND CHANGES, 1949–1969

Increase
31 - 46.9%
15 - 30.9%
0 - 14.9%
Decrease
-13.1%
0 10 20 30 40 Miles
0 50 Kilometers

64. ROW CROPS SINCE 1950

64. ROW CROPS SINCE 1950

Since 1950, Arkansas agriculture has experienced major changes in the row crops emphasized. The emergence of alternative cash crops, stiffer competition from competitors in California and Texas, and reduced demand have all encouraged farmers to abandon cotton for other crops. Between 1949 and 1970 there was a steady decline in the importance of cotton. After production reached an all-time high of 1,587,307 bales in 1950 the reported harvest declined to 1,535,750 bales in 1964 and 852,309 bales in 1974. By 1980 production had declined to 440,000 bales.

The soybean replaced cotton as the most important Arkansas row crop. In 1950 farmers harvested 6,134,516 bushels of beans, and subsequent production grew steadily. In 1964 the total production was 64,669,500 bushels, and by 1974 it had reached 76,365,435 bushels. Encouraged by the expansion of overseas markets for the product, much of the farmland of eastern Arkansas has been turned over to its cultivation.

Rice has also been an important crop and has been alternated with soybeans on eastern farms. In 1950, Arkansas farmers harvested 19,889,614 bushels of rice. In 1980 the state was the leading rice producer in the nation, harvesting 52,615,000 hundredweight. Like soybeans, the principal market for this product has been overseas.

Arkansas farmers have continued to emphasize commercial agriculture, but since 1950 the relative importance of each crop has changed. In 1950 cotton produced 46.6 percent of the total cash receipts in agriculture; rice, 8.4 percent; and soybeans, 4.7 percent. By 1960 cotton's share had declined to 33.9 percent, rice had increased to 9.1 percent, and soybeans had leaped to 14.7 percent. The 1960's brought about more marked changes: in 1970, at the end of the decade, cotton accounted for only 11.3 percent of revenues; rice had yielded 9.8 percent; and soybeans, 26.9 percent. This basic pattern has persisted into the 1980's, although the combined share of all three crops has become less important when compared with livestock, especially poultry production. In 1983 the relative shares of agricultural receipts from the three crops were cotton, 6.7 percent; rice, 10.8 percent; and soybeans, 21.4 percent.

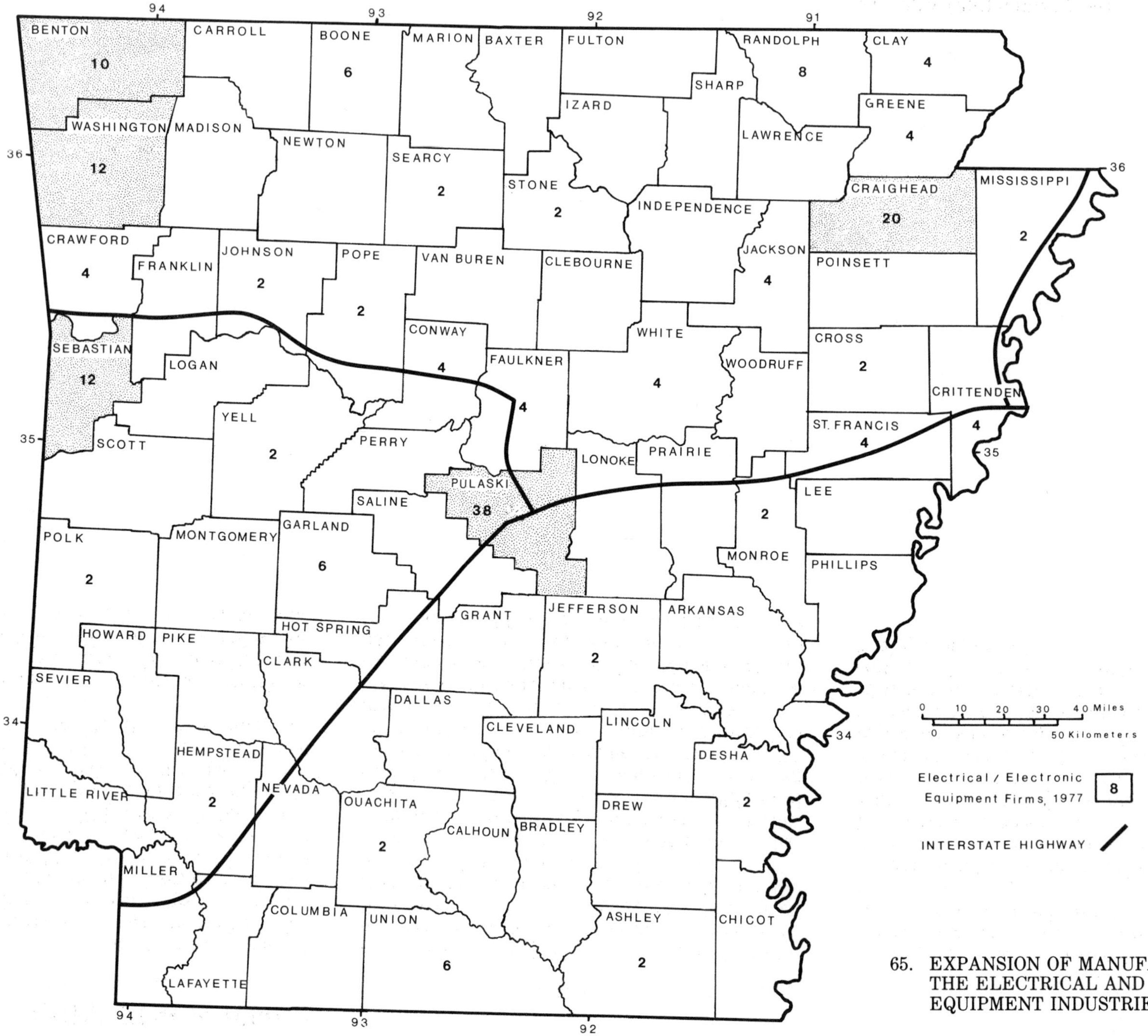

65. EXPANSION OF MANUFACTURING: THE ELECTRICAL AND ELECTRONIC EQUIPMENT INDUSTRIES

65. EXPANSION OF MANUFACTURING: THE ELECTRICAL AND ELECTRONIC EQUIPMENT INDUSTRIES

The reliance on agriculture and extractive industries produced major problems in the Arkansas economy during the Second World War. As a result, business and political leaders began in the mid-1940's to find some means of diversifying the state's economy. Their efforts, plus general changes in manufacturing in the United States, produced a greater degree of diversification and, as a result, a general expansion of economic opportunities in the state in the last thirty years.

The efforts taken in Arkansas to bring about economic diversification came as a result of the organization of the Arkansas Economic Council in 1943. Along with the Arkansas Chamber of Commerce, business leaders on the council urged public officials to work to bring in new industries and to plan the state's economic development. A major result of their efforts was the creation in 1945 of the Arkansas Resources and Development Commission, which emphasized the introduction of manufacturing industries into the state in order to improve economic conditions. In 1955 the administration of Governor Orval Faubus moved to encourage industrialization by the creation of the Arkansas Industrial Development Commission, or AIDC. Headed by Winthrop Rockefeller, the AIDC was charged with soliciting new industries, expanding existing industries, and providing whatever aid was necessary to accomplish these goals. The AIDC legislation also allowed the creation of local industrial development corporations to assist new industries. Following the creation of the AIDC, the General Assembly passed other acts to encourage industrial expansion. In 1957 it created the First Arkansas Development Finance Corporation to provide loans to industrial groups. Also in 1957, the state government, with the University of Arkansas, established the Industrial Research and Extension Center and the Graduate Institute of Technology to provide technical and scientific support for new industries. Amendment 49, passed in 1958, allowed local governments to issue bonds for industrial development. In 1960, Act 9 gave local authorities further powers to support new industries, including the power to buy and equip such properties.

As a result of this active pursuit of industrial operations, the manufacturing component of Arkansas's economy rapidly expanded after 1950. The AIDC reported that it helped some three thousand new industries develop in the state between 1955 and 1972.

Food processing, aluminum, and lumbering remained major parts of the manufacturing economy, but new companies also produced clothing, shoes, electrical machinery, and other products. Unfortunately, many of these industries relied on unskilled labor, so that, while industrial employment improved, wages did not rise rapidly. In addition, in the last decade many of the operations performed in Arkansas have been transferred overseas to take advantage of even cheaper labor.

The most important new industry, besides those connected with food processing, aluminum, and timber, has been the manufacture of electrical and electronic products, which has developed throughout the state with the modern highway system. The accompanying map shows how this important industry has emerged along the major transport routes shown in Map 62.

65. EXPANSION OF MANUFACTURING:
THE ELECTRICAL AND ELECTRONIC
EQUIPMENT INDUSTRIES

1940–1960

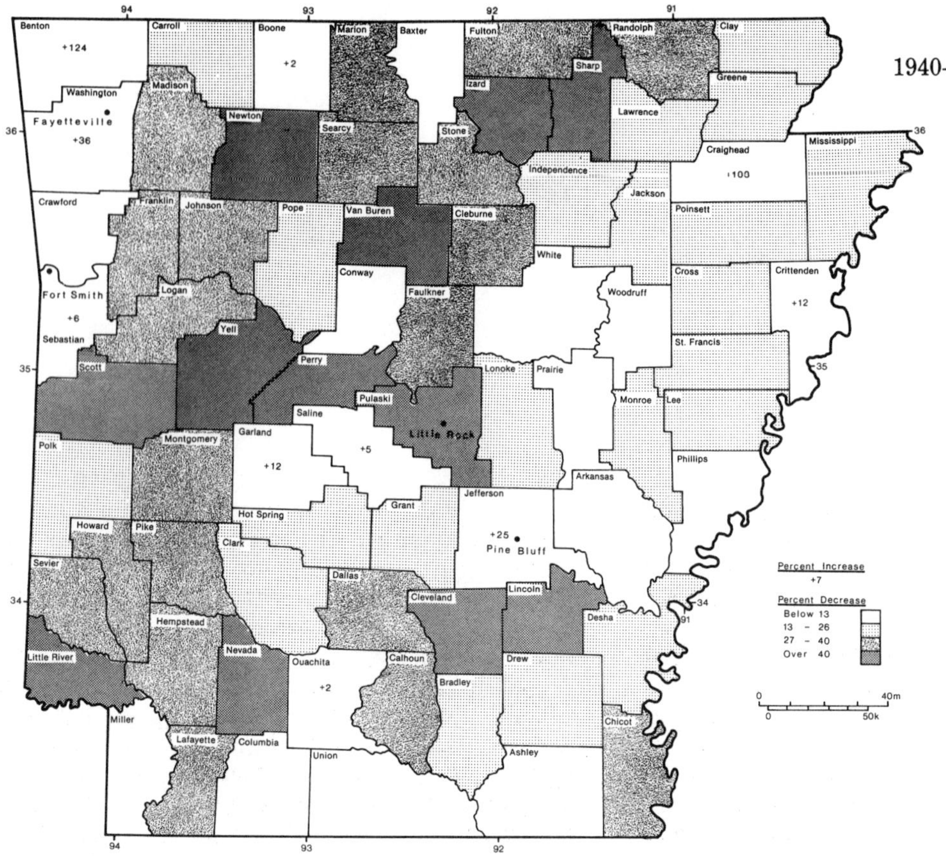

1960–1980

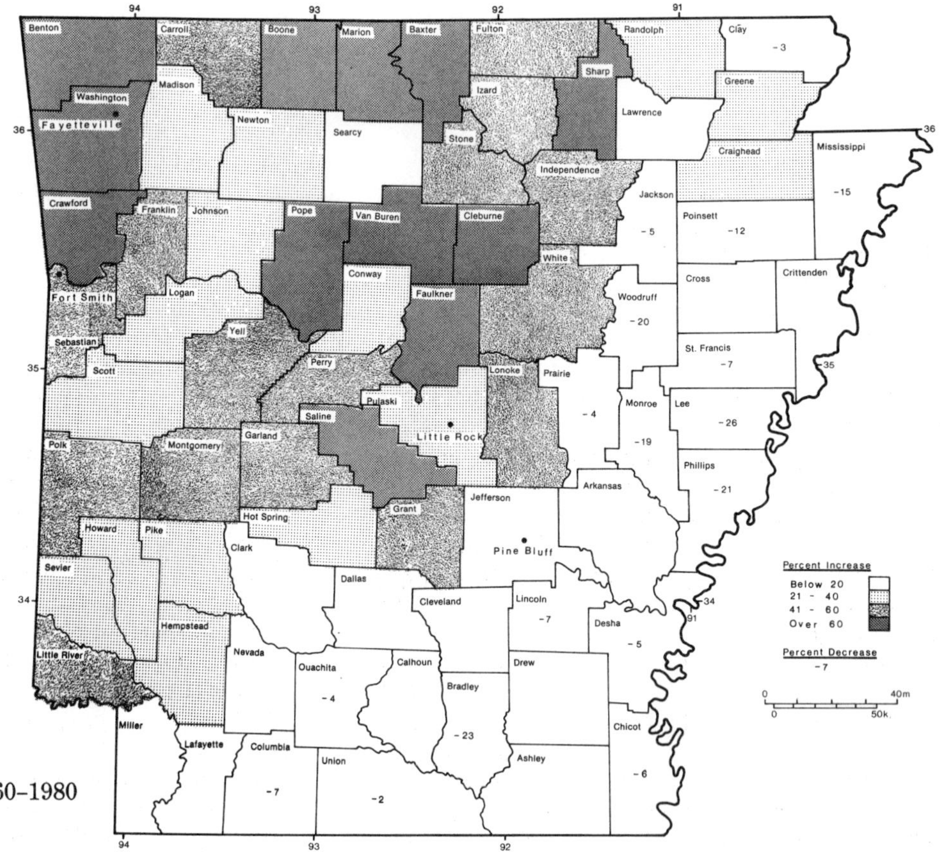

66. GENERAL POPULATION GROWTH SINCE 1950

66. GENERAL POPULATION GROWTH SINCE 1950

Population in Arkansas continued to decline in the 1950's, but since 1960 it has shown steady growth. Economic diversification has created new opportunities that attract outsiders to the state and keep the state's young people at home. Arkansas has benefitted from being in the nation's Sun Belt, although not as much as other southern states. The development of the northwestern section of the state as a retirement center has also brought additional people into Arkansas.

By 1950 the state's population had reached 1,909,511, but then it dropped to 1,786,272 in 1960. This decline reflected the continuing agricultural problems of the time. The loss of population continued to be greatest in the rural areas, although all areas suffered. The population of 1960 turned out to be the low point, however, and since then population growth has been strong.

The 1960s brought general population growth, and by 1970 the total population had climbed beyond that of 1950, to 1,923,295. The greatest growth rates were in the northwestern counties, which benefitted from the expansion of the poultry industry, some electronic manufacturing, and the influx of retirees. Baxter County led the state with a growth rate of 54.1 percent, and was followed by Benton, with 39.2 percent, and Washington, with 38.7 percent.

In 1980 the population of Arkansas was 2,298,000. The growth rate, 19.5 percent, was at a level not achieved since the 1910's. The greatest increases continued to be in the northwestern and central sections of the state. Reflecting the emergence of opportunities related to manufacturing, much of the growth was in the cities. By 1980 the majority of Arkansans, 51.6 percent, lived in urban areas.

Population of Arkansas

Year	Total	Percent Increase or Decrease
1950	1,909,511	
1960	1,786,272	-6.5
1970	1,923,295	7.7
1980	2,298,000	19.5
1983	2,328,000	1.3
1990 (proj.)	2,579,800	12.6

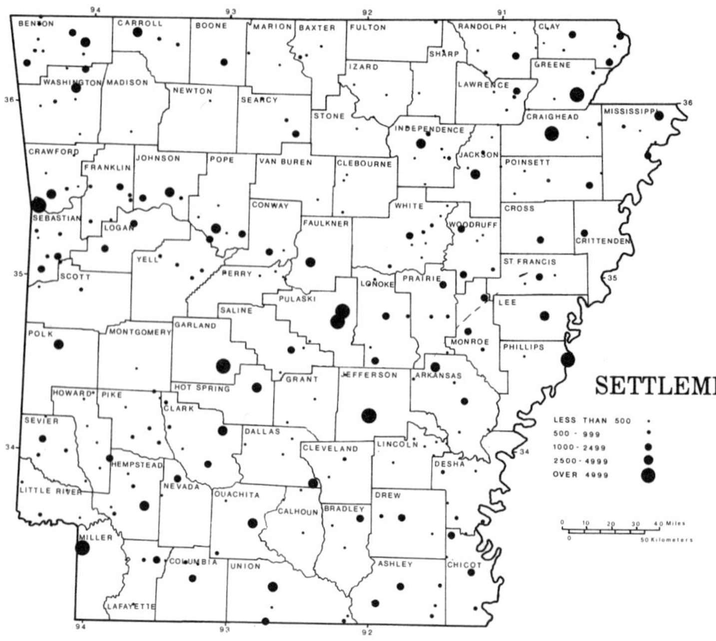

SETTLEMENT IN 1910

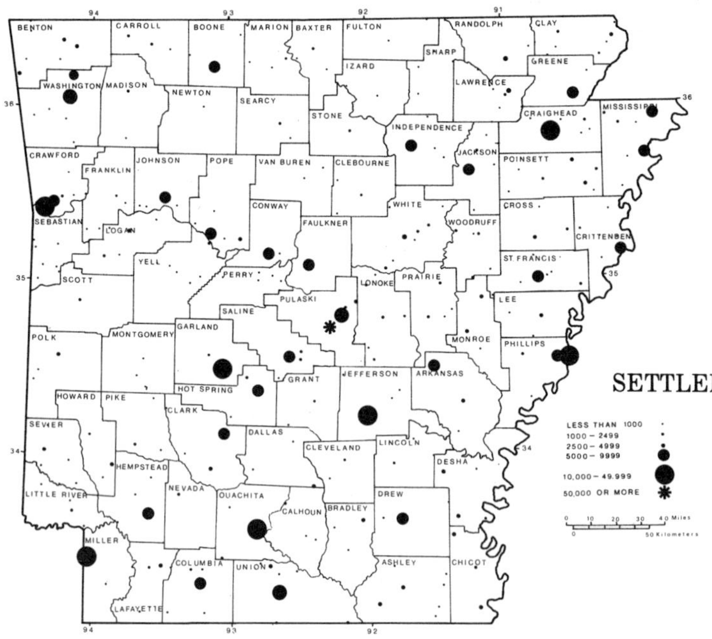

SETTLEMENT IN 1950

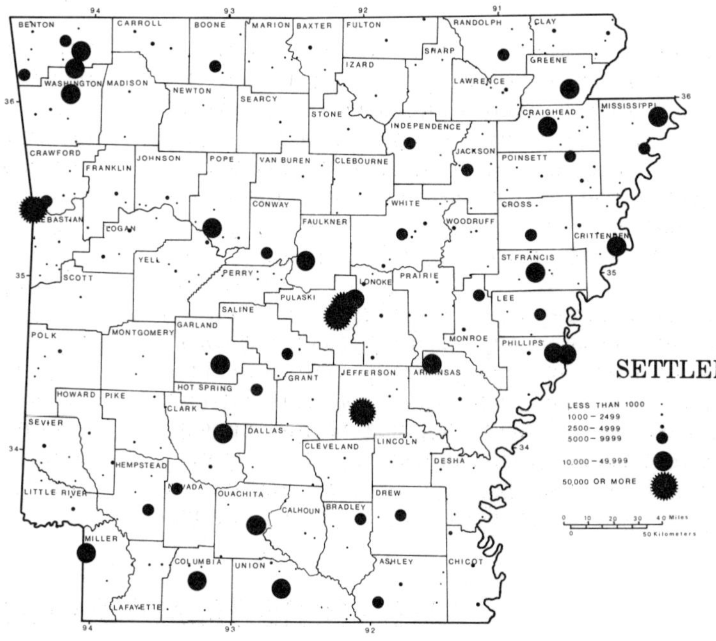

SETTLEMENT IN 1970

67. URBANIZATION 1900–1980

67. URBANIZATION, 1900–1980

The most important demographic trend in Arkansas during the first eight decades of the twentieth century has been the movement of people from the countryside into towns, a process known as urbanization. Declining opportunities in agriculture pushed people off the farm through much of this period. While the general population figures showed that parts of that population left the state, a relatively prosperous urban economy drew many to the state's towns. A population that was primarily rural in 1900 has changed to a population of which half were urban in 1970, and over half were urban in 1980.

The United States Bureau of the Census defines an urban community as an incorporated area with a population in excess of 2,500 persons. Using that definition, the urban population of the state in 1900 was 111,733, a little over 8 percent of the whole population. By 1920 the urban population had grown to 290,497, or 20 percent. This basic trend has continued to the present, as shown by the following figures: 382,878, or 20.6 percent, in 1930; 431,910, or 22.1 percent, in 1940; 617,153, or 32 percent, in 1950; 765,303, or 43 percent, in 1960; 878,624, or 50 percent, in 1970; and 1,178,349, or 51.5 percent, in 1980.

The trend was particularly obvious during the first three decades of the twentieth century, when the rate of urban population growth was 243 percent, compared to the general population expansion of 41 percent. Between 1940 and 1960 a similar urban explosion happened when cities grew at a rate of 77 percent, compared to an 8 percent decline in the general population. At other times urban growth has been lower than these periods, but it has never been less than twice as great as the growth rate of the general population.

The maps show the general pattern of urban growth by presenting incorporated places having a population of 5,000 and more. In 1900 only five towns had such populations. These centers of trade and commerce for the state were Little Rock, in Pulaski County, 38,307; Fort Smith, in Sebastian County, 11,587; Pine Bluff, in Jefferson County, 11,496; Hot Springs, in Garland County, 9,973; and Helena, in Phillips County, 5,550. During the next ten years (see Map 67a) this number increased to nine with the addition of Jonesboro, in Craighead County; North Little Rock, in Pulaski County; Paragould, in Greene County; and Texarkana, in Miller County.

By 1950 (see Map 67b) the number of towns with a population of 5,000 or more had grown to thirty-one. Little Rock and North Little Rock together continued to be the largest metropolitan area with 146,310 people. Fort Smith remained second with 47,942, and Pine Bluff third with 37,162.

By 1970 there were forty-two towns with 5,000 or more residents. The Little Rock–North Little Rock population had reached 192,523. Other early leaders continued to grow. Fort Smith continued to be second with 62,802, and Pine Bluff remained third with 57,389. There was a general spread of urban culture throughout the state, and local commercial and administrative centers, especially county seats, were expanding. In 1970 the towns of northwest Arkansas were showing particular growth. The towns of Bentonville, Fayetteville, Rogers, and Springdale, located in Benton and Washington counties, had together a population of 64,070, making the combined metropolitan area the third largest in the state. The growth rates of Fayetteville, Rogers, and Springdale were all in excess of 50 percent.

By 1980 fifty towns in Arkansas had a population of five thousand or more. The major commercial centers continued to expand, while rural communities continued to grow.

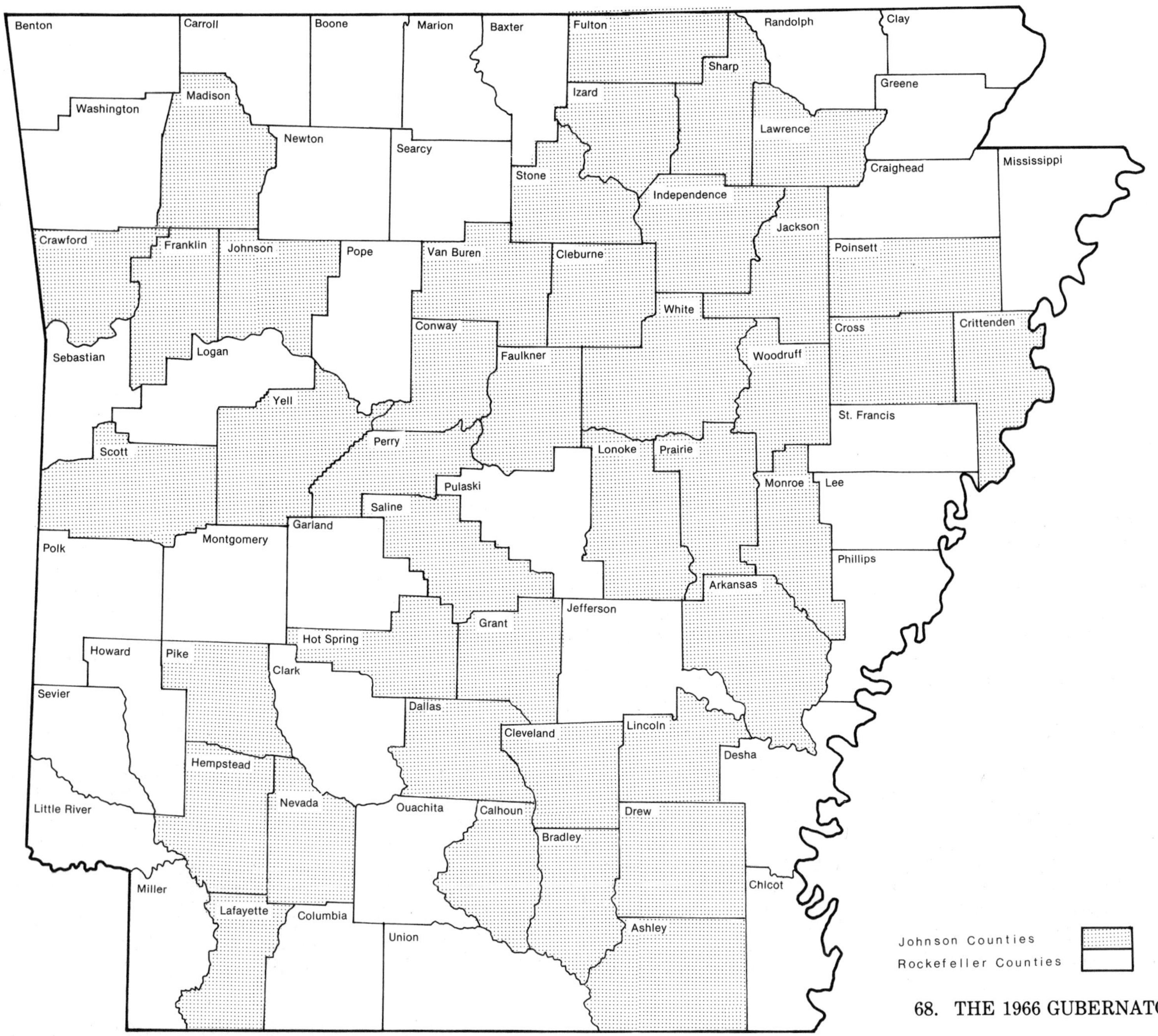

68. THE 1966 GUBERNATORIAL ELECTION

68. THE 1966 GUBERNATORIAL ELECTION

Changes in the basic structure of the state's economy and society have resulted in considerable upheaval in Arkansas politics. Most noteworthy has been the weakening of the Democratic party's domination of state government. While the Democratic party can no longer count on the supporters and the power that it possessed for one hundred years after the Civil War, new voting patterns have not stabilized. Politics in more recent years have seen constantly changing coalition patterns.

The first major challenge to traditional Democratic rule at the state level came in the gubernatorial election of 1966. In that year the Republican party nominated Winthrop Rockefeller as its candidate for governor. Rockefeller, youngest son of John D. Rockefeller, had moved in 1953 to Arkansas, where he set up farm and ranching activities. He was brought into politics in 1955 when Governor Orval Faubus named him to head the Arkansas Industrial Development Commission. In 1964 the Republicans nominated him for governor for the first time. Although he lost in his first race, Rockefeller secured 43 percent of the vote.

In 1966 the Republicans nominated Rockefeller again, and the Democratic party selected as its candidate Justice Jim Johnson. Johnson ran a traditional Democratic campaign that emphasized opposition to change, particularly with regard to the state's segregationist policies. Rockefeller responded with a campaign that emphasized economic development, repudiated racism, and attacked the power of the Democratic establishment, alleging wrong doing in the state highway commission and the game and fish commission.

In the campaign Rockefeller managed to put together a new coalition. His support came from a variety of sources that in other instances might have been quite opposed to each other. Urban whites, seeking economic opportunity, and blacks throughout the state, who saw Rockefeller as preferable to Johnson, joined with traditional Republican voters in northwestern Arkansas to produce a Rockefeller majority. He won 299,105 votes to Johnson's 254,338.

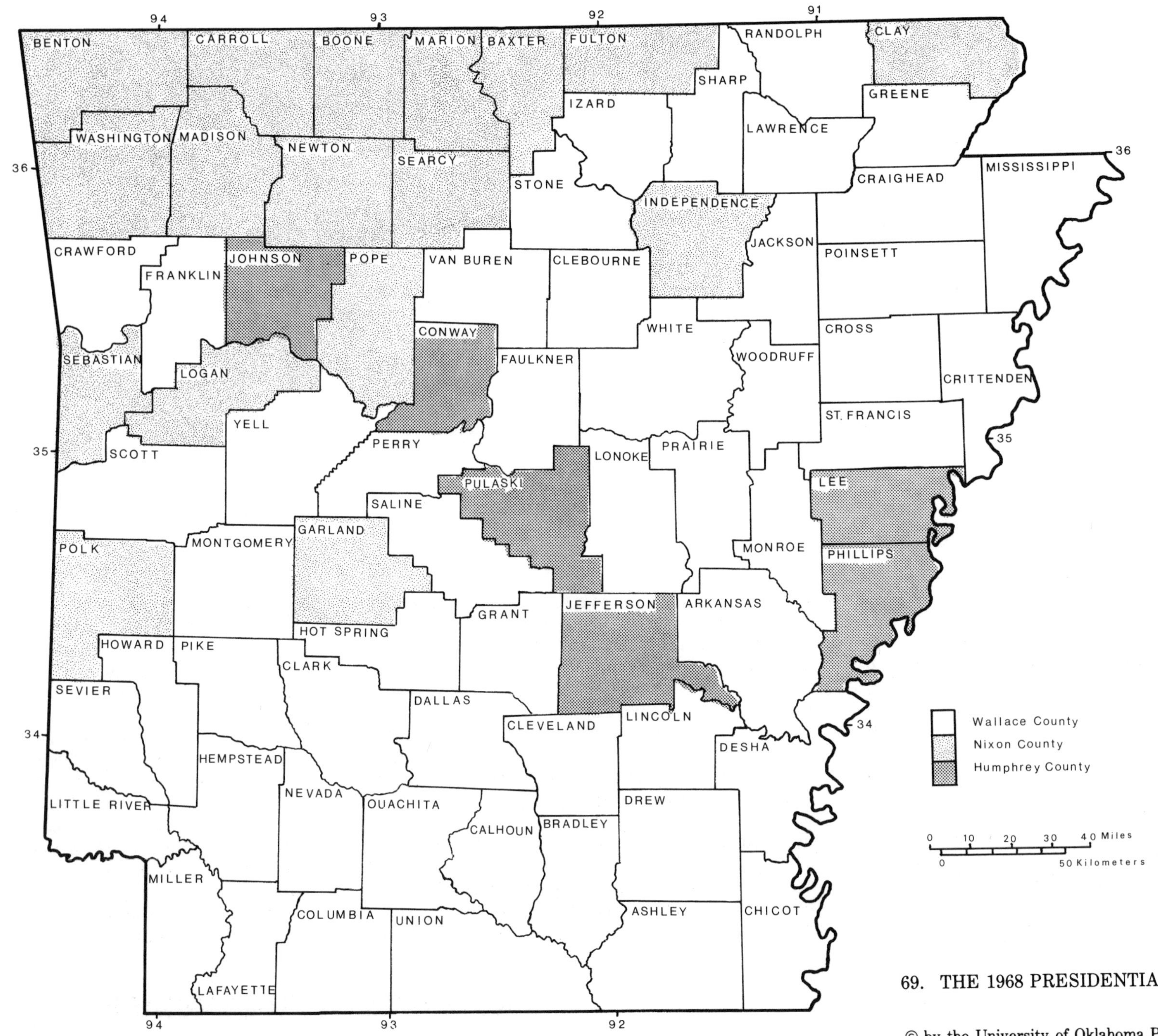

69. THE 1968 PRESIDENTIAL ELECTION

69. THE 1968 PRESIDENTIAL ELECTION

Only two years after Winthrop Rockefeller's election as governor, the national elections showed that party labels were making less of difference to modern Arkansans. Coalitions were in the process of constantly changing, and another major pattern emerged when George C. Wallace ran as a third-party candidate for president.

The 1968 election took place at a time of considerable social and political upheaval. The year was marked by the assassination of the Reverend Martin Luther King, whose death sparked rioting in over sixty American cities. Robert Kennedy, brother of former President John F. Kennedy and a major candidate for the Democratic presidential nomination, was assassinated in June. The year was also marked by protests and opposition to the war in Viet Nam.

The Democratic party was in turmoil at its national convention, an event marked by protest and riot. The party's candidate, Hubert H. Humphrey, had been Lyndon B. Johnson's vice-president, and was too closely linked with the Johnson administration's war policy to be acceptable to many of the more liberal members of the party, and too closely associated with civil rights reform for many who were more conservative. He was opposed by the Republican Richard M. Nixon. For many white Arkansans a viable third-party alternative emerged with the candidacy of George Wallace on the ticket of the American Independant party.

In the national election Nixon won handily, receiving 31,785,480 votes, 43.4 percent of the total, to Humphrey's 31,275,188, or 42.7 percent, and Wallace's 9,906,473, or 13.5 percent. Wallace's candidacy deprived Humphrey of the electoral votes of five southern states, including Arkansas, and possibly secured the electoral votes of other southern states for Nixon by draining away traditional Democratic voters. In Arkansas, Wallace received 235,627 votes, or 38.7 percent, of the total to Humphrey's 184,901 and Nixon's 189,062.

In the Arkansas election the traditional regional splits appeared to be important. The Republicans carried most of the far northwestern counties, indicating the permanency of the Republican pattern there. Garland County also showed its growing tendency towards Republican support. The vote showed, however, that some of the western voters were more attracted to conservative causes than the Republican party because they gave their support to Wallace. Humphrey carried only the state's urban center, Pulaski County; three counties with large black populations, Jefferson, Lee, and Phillips; and Johnson and Conway counties. Wallace showed strength throughout the state.

In addition to the trends shown by the presidential election, Arkansans demonstrated a considerable degree of discrimination in their choices. While voting for Wallace, they also returned Senator J. William Fulbright to the United States Senate over a Republican opponent. In the state gubernatorial election they reelected Winthrop Rockefeller as governor, who, when first elected in 1966, had been the first Republican chosen for that office in the state since Reconstruction.

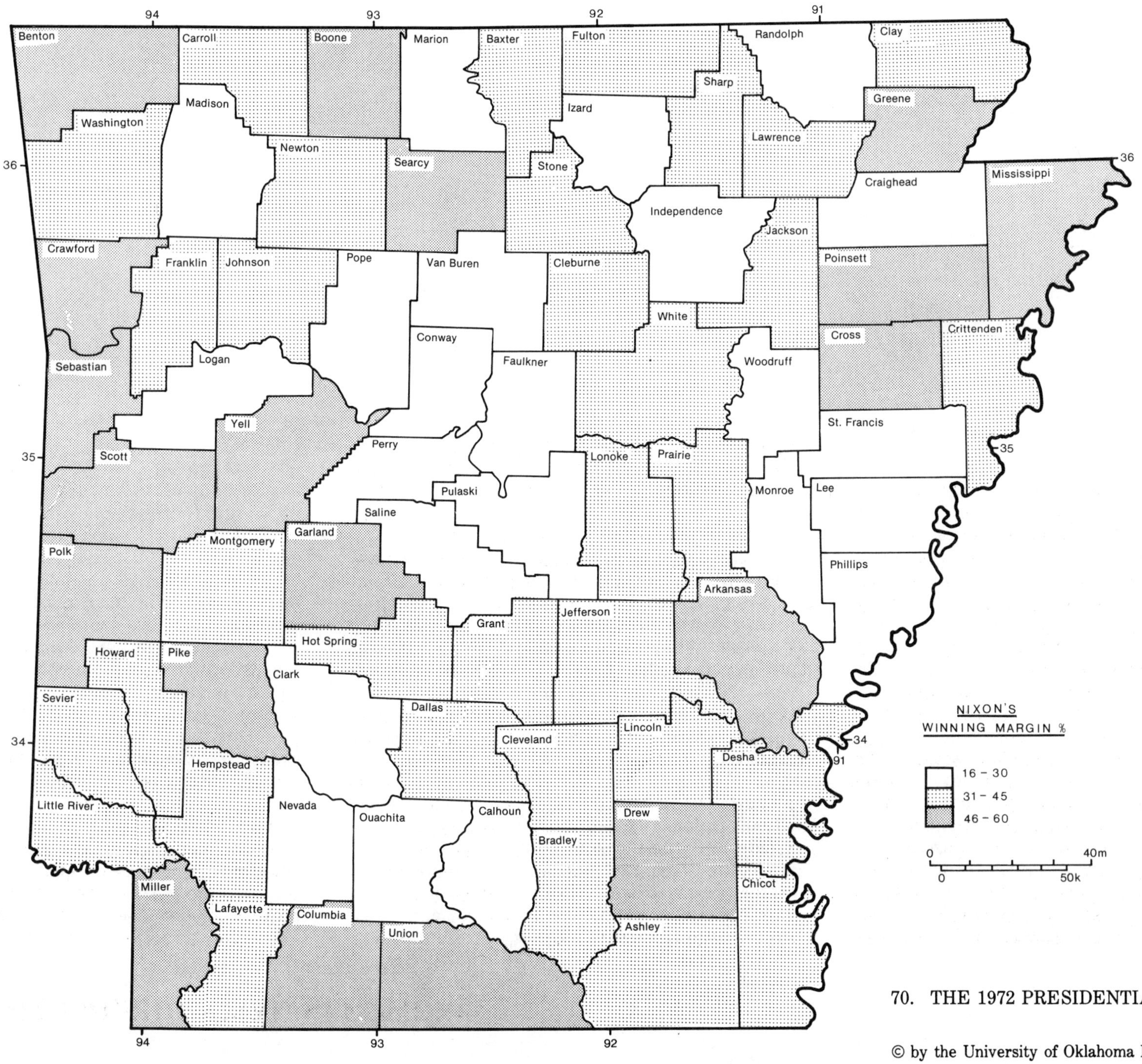

70. THE 1972 PRESIDENTIAL ELECTION

70. THE 1972 PRESIDENTIAL ELECTION

In 1972, Richard M. Nixon was the first Republican presidential candidate to receive a majority of the votes in Arkansas since the election of Ulysses S. Grant during Reconstruction in 1872. Nixon appealed particularly to white voters, who were, in part, attracted to his opposition to the civil rights legislation developed during the Kennedy and Johnson years. Although he was unsuccessful in his efforts to block renewal of the Voting Rights Act in 1970, and he failed to prevent busing as a desegregation measure in urban areas, Nixon's position made him popular among those with similar views. As president his foreign-policy triumphs—the detente with the Soviets, the recognition of Communist China, and winding down the war in Viet Nam—also attracted many voters. His chances for reelection were threatened, however, by economic problems: the country was experiencing inflation sparked by rising energy prices as industrial productivity began to turn down.

Nixon's opponents were never able to challenge the president in the election. George Wallace, who was as a third-party candidate might have drawn conservative supporters away from the president, was shot in May, 1972; left permanently paralyzed, he was forced to withdraw from the campaign. The Democratic campaign of George S. McGovern never got off the ground. Traditional Democratic support failed to materialize from the urban political machines of the North, and the AFL–CIO, another traditional ally, refused to endorse McGovern's candidacy. McGovern's apparent representation of the liberal wing of the Democratic party did not secure him much support among the white majority in Arkansas.

In the election Nixon received 60.7 percent of the popluar vote nationally, 47,169,911 to McGovern's 29,170,383. In Arkansas, Nixon received 445,751 votes to McGovern's 198,899. His total was 69 percent of the vote, more than the national triumph that he achieved. The election was even more remarkable because Nixon carried every county in the state. His greatest pockets of strength were in the western and southwestern counties, but he also did very well in traditionally Democratic counties in southeastern Arkansas and along the Mississippi.

As would be shown in 1976, when Democratic candidate Jimmy Carter defeated Republican Gerald R. Ford, the triumph of 1972 appears to have been a personal one for President Nixon and not the result of any major party realignments in Arkansas. Arkansans returned to their tradition of supporting Democratic candidates in 1976.

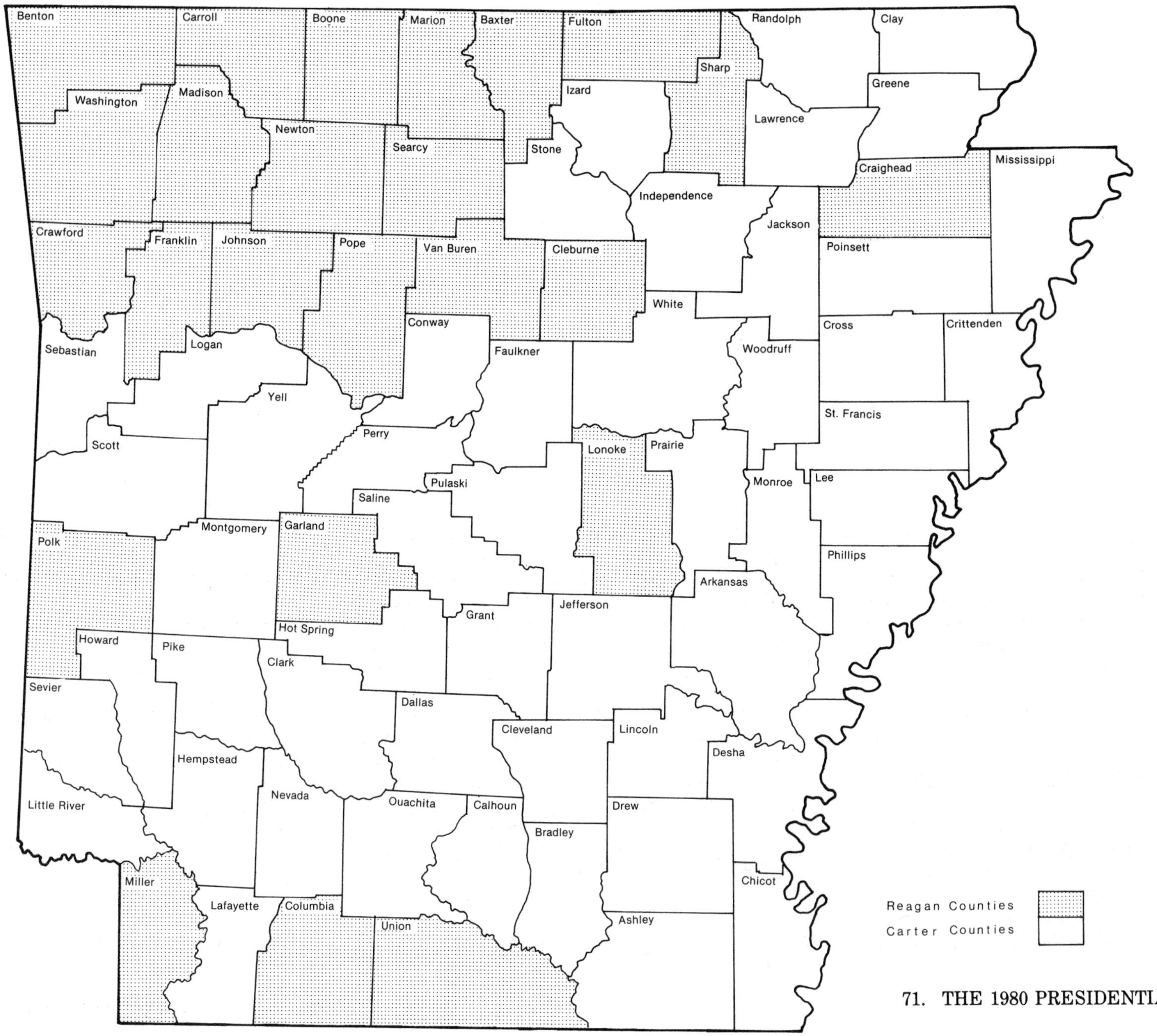

71. THE 1980 PRESIDENTIAL ELECTION

71. THE 1980 PRESIDENTIAL ELECTION

In 1976, Arkansas voted with its historical Democratic loyalty, but in 1980 it joined the national majority in voting for Ronald Reagan. Whether or not this shift indicated a more permanent realignment of Arkansas voters is difficult to assess.

Reagan ran in 1980 as a defender of traditional and local values. He was vocal in his support of traditional family life, prayer in schools, tough criminal laws, and a strong military. Nationally, Reagan defeated his opponent, the incumbent Jimmy Carter, with about 44 million votes to only 35 million. His electoral vote was 489 to 49. His 403,164 votes in Arkansas were enough to give him 48 percent of the vote and a plurality of 4,195 votes.

Several factors help to explain Reagan's victory in Arkansas. The increasing percentage of the population classified as elderly has been considered by some experts an important factor in Reagan's victory nationally. If people in this demographic category voted as they did in 1980 because they were ideologically more in tune with Reagan and his party, Arkansas was in line for such an appeal. Reagan did particularly well in counties with large populations of retired persons in northwestern Arkansas. He also carried Garland County with its similarly constituted population. The changing age of the state's population could bring about a permanent realignment of traditional political loyalties in the future, but it is too early to tell if that will be true.

Other important factors in the election, however, make it more difficult to establish a pattern based on age or any other single determinant. In the 1980 election Reagan's opponent suffered from frustrations throughout society generated by the Iranian hostage crisis. High rates of inflation further weakened Carter by ravaging the domestic economy. In the days before the Democratic national convention, Carter had lost the support of elements of his own party as a result of Edward M. Kennedy's attempt to gain the party nomination. It may well be that the election reflected only a growing tendency in the nation to emphasize the personality of individual politicians, and that conditions in Arkansas in 1980 made it impossible for Carter to convince the electorate that he could continue to lead the nation.

BIBLIOGRAPHY

GENERAL BOOKS ON ARKANSAS

Ashmore, Harry S. *Arkansas: A Bicentennial History.* New York: W.W. Norton & Co., 1976.

Dougan, Michael. *Confederate Arkansas.* Tuscaloosa: University of Alabama Press, 1976.

Du Vall, LeLand. *Arkansas: Colony and State.* Little Rock: Rose Publishing Co., 1973.

Ferguson, John L., and James H. Atkinson. *Historic Arkansas.* Little Rock: Democrat Printing and Lithographing Co., 1966

Herndon, Dallas T., ed. *Centennial History of Arkansas.* 3 vols. Chicago: S. J. Clarke Publishing Co., 1922

Moore, Waddy W. *Arkansas in the Gilded Age, 1874–1900.* Little Rock: Rose Publishing Co., 1976.

Thompson, George. *Arkansas and Reconstruction: The Influences of Geography, Economics and Personality.* Port Washington, N.Y.; Kennikat Press, 1976.

Wagner, Janice, ed. *The Historical Report of the Secretary of State.* Vol. 3. Little Rock: Winston Bryant, Arkansas Secretary of State, 1978.

Williams, C. Fred, et al. *A Documentary History of Arkansas.* Fayetteville: University of Arkansas Press, 1984.

MAPS 1 AND 2. LOCATION AND NATURAL REGIONS

Patterson, J. H. *North America.* 6th ed. New York: Oxford University Press 1982.

Pirkle, E. C., and William Yoho. *The Natural Regions of the United States.* 2d ed. Dubuque, Iowa: Kendall-Hunt, 1977.

Stroud, H. B., and Gerald T. Hanson. *Arkansas Geography: The Physical Landscape and the Historic-Cultural Setting.* Little Rock: Rose Publishing Company, 1981.

MAP 3. SOIL PATTERNS

Crow, Charles. *Arkansas Natural Area Plan.* Little Rock: Arkansas Department of Planning, 1974.

MAP 4. SURFACE WATER BODIES

Jackson, Jeanne, and Leslie E. Mack. *Arkansas Water: Why Wait for the Crisis.* Little Rock: Winthrop Rockefeller Foundation, 1982.

MAPS 5, 6, AND 7. TEMPERATURE AND PRECIPITATION PATTERNS AND GROWING SEASONS

Climates of the States. Detroit: Gale Research Company, 1973.

Skiles, Albert. *Climatic Atlas of Arkansas.* Little Rock: Arkansas Department of Energy, 1981.

MAP 8. VEGETATION PATTERNS

Crow, Charles. *Arkansas Natural Area Plan.* Little Rock: Arkansas Department of Planning, 1974.

Nelson, Thomas, and Walter Zillgit. *A Forest Atlas of the South.* New Orleans: United States Forest Service, 1969.

MAPS 9, 10, 11, AND 12. INDIAN SETTLEMENTS FROM BEFORE 8,000 B.C. TO 1500 A.D.

Chapman, Carl. "Cultural Sequence in the Lower Missouri Valley." In *Archaeology of the Eastern United States,* edited by James Griffin, 139–52. Chicago: University of Chicago Press, 1952.

Griffin, James. "Prehistoric Cultures of the Central Mississippi Valley." In *Archaeology of the Eastern United States,* edited by James Griffin, 226–38. Chicago: University of Chicago Press, 1952.

Hoffman, Michael P. "Prehistoric Developments in Southwestern Arkansas." *Arkansas Archaeologist* 10 (1969): 37–44.

Kneberg, Madeline. *"The Tennessee Area."* In *Archaeology of the Eastern United States,* edited by James Griffin. 190–206. Chicago: University of Chicago Press, 1957.

McClurkan, Burney B. "Culture History of Southeast Arkansas." *Arkansas Archaeologist* 10 (1969): 31–34.

McGimsey, Charles R., III. *Indians of Arkansas.* Series, no. 1. Fayetteville: Arkansas Archaeological Survey, 1969.

Morse, Dan F. *Archaeology of the Central Mississippi Valley.* New York: Academic Press, 1983.

______. "Introducing Northeastern Arkansas Prehistory." *Arkansas Archaeologist* 10 (1969) : 13–23.

Scholtz, James A. "A Survey of Prehistory in Northwest Arkansas." *Arkansas Archaeologist* 10 (1969) : 24–30.

Trubowitz, Neal, and Marvin Jeter, eds. *Arkansas Archaeology in Review.* Research Series, no. 15. Fayetteville: Arkansas Archaeological Survey, 1982.

MAPS 13, 14, AND 15. INDIAN SETTLEMENT, 1500–1800 A.D.

Baird, W. David. *The Quapaw Indians: A History of the Downstream People.* Norman: University of Oklahoma Press, 1980.

Dickinson, Samuel Dorris. "Lake Mitchegamas and the St. Francis." *Arkansas Historical Quarterly* 43 (Autumn, 1984): 197–207.

McGimsey, Charles R., III. *Indians of Arkansas.* Popular Series, no. 1. Fayetteville: Arkansas Archaelogical Survey, 1969..

Swanton, John R. *The Indian Tribes of North America.* Pp. 212–13. Smithsonian Institution, Bureau of American Ethnology Bulletin no. 145. Washington, D.C.: United States Government Printing Office, 1952.

MAP 16. INDIAN TRIBES IN PROXIMITY TO ARKANSAS

Din, Gilbert C. "The Spanish Fort on the Arkansas, 1763–1803." *Arkansas Historical Quarterly* 42 (Autumn, 1983): 271–93.

Kroeber, Alfred L. *Culture and Natural Areas of Native North America.* University of California Publications in American Archaeology and Ethnology, no. 38. Berkley: University of California, 1939.

Swanton, John R. *The Indian Tribes of North America.* Washington D.C.: United States Government Printing Office, 1952.

______. *The Indians of the Southeastern United States.* Reprint, New York: Greenwood Press, 1969.

MAPS 17 AND 18. INDIAN MIGRATION IN ARKANSAS, 1700–1800 AND 1800–1836

Dale, Edward E. "Arkansas and the Cherokees." *Arkansas Historical Quarterly* 8 (Summer, 1949): 95–114.

Swanton, John R. *The Indian Tribes of North America.* Washington D.C.: United States Government Printing Office, 1952.

______. *The Indians of the Southeastern United States.* Reprint, New York: Greenwood Press, 1969.

MAPS 19 AND 20. INDIAN LAND CESSIONS, 1803-1818 AND 1818-1835

Baird, W. David. "Arkansas Choctaw Boundary: A Study of Justice Delayed." *Arkansas Historical Quarterly* 28 (Autumn, 1969): 203-22.

Gabler, Ina. "Lovely's Purchase and Lovely County." *Arkansas Historical Quarterly* 19 (Spring, 1960): 31 45.

Hempstead, Fay. *Historical Review of Arkansas: Its Commerce, Industry and Modern Affairs.* 1:55-71. Chicago: Lewis Publishing Company, 1911.

Lane, Jack. "Federal-Quapaw Relations, 1800-1833." *Arkansas Historical Quarterly* 19 (Spring, 1960): 61-74.

White, Lonnie J. "Arkansas Territorial Indian Affairs." *Arkansas Historical Quarterly* 21 (Autumn, 1961): 193-212.

MAP 21. DE SOTO'S JOURNEY, 1541-1542

Final Report of the United States De Soto Expedition Commission. Pp. 226-576. Washington D.C.: United States Government Printing Office, 1939.

Herndon, Dallas T., ed. *Centennial History of Arkansas.* 1:73-84. Chicago and Little Rock: S. J. Clarke Publishing Company, 1922.

Hudson, Charles, Jr. "De Soto in Arkansas: A Brief Synopsis." *Fieldnotes* (Arkansas Archaeological Society newsletter), no. 205 (1985), 3-12.

Maynard, Theodore. *De Soto and the Conquistadores.* London: Longmans, Green and Company, 1930.

"The Narrative of the Expedition of Hernando De Soto, By the Gentleman of Elvas." In *Original Narratives of Early American History,* edited by J. Franklin Jameson. New York: Charles Scribner's Sons.

MAP 22. FRENCH EXPLORATION, 1673-1740

Cox, Isaac Joslin, ed. *The Journeys of Rene Robert Cavelier, Sieur de La Salle.* New York: A.S. Barnes & Company, 1905.

Hempstead, Fay. *Historical Review of Arkansas: Its Commerce, Industry, and Modern Affairs.* 1:4-5, 9-12, 18. Chicago: Lewis Publishing Company, 1911.

Herndon, Dallas T., ed. *Centennial History of Arkansas.* 1:85, 90. Chicago and Little Rock; S. J. Clark Publishing Company, 1922.

Smith, Ralph A., ed. "Exploration of the Arkansas River by Bernard de la Harpe, 1721-1722." *Arkansas Historical Quarterly* 10 (Winter, 1951): 339-63.

MAP 23. AMERICAN EXPLORATION, 1803-1819

Allen, Milford A. "Thomas Jefferson and the Louisiana Frontier." *Arkansas Historical Quarterly* 20 (Spring, 1961): 39-64.

Beidleman, Richard G. "The 1818-20 Arkansas Journey of Thomas Nuttall." *Arkansas Historical Quarterly* 15 (Autumn, 1956): 249-59.

Dunbar, William. *Documents Relating to the Purchase and Exploration of Louisiana.* Boston: Houghton Mifflin & Co., 1909.

Featherstonhaugh, George W. *Excursion Through the Slave States.* New York: Harper and Brothers, 1844.

Hempstead, Fay. *Historical Review of Arkansas: Its Commerce, Industry, and Modern Affairs.* 1:49-50. Chicago: Lewis Publishing Company, 1911.

Herndon, Dallas T., ed. *Centennial History of Arkansas.* 1:603-604. Chicago: S. J. Clark Publishing Company, 1922.

Nuttall, Thomas. *Journal of Travels into the Arkansas Territory During the Year 1819.* Edited by Savoie Lottinville. Norman: University of Oklahoma Press, 1980.

Schoolcraft, Henry Rowe. *Schoolcraft in the Ozarks: Report of Journal of a Tour into the Interior of Missouri and Arkansas.* Edited by Hugh Park. Van Buren, Ark.: Argus Press, 1955.

MAP 24. FRENCH LAND SUBDIVISION

Arnold, Morris S. *Unequal Laws unto a Savage Race: European Legal Traditions in Arkansas, 1686-1836.* Fayetteville: University of Arkansas Press, 1985.

Herndon, Dallas T., ed. *Centennial History of Arkansas.* 1:93-98. Chicago and Little Rock: S. J. Clarke Publishing Company, 1922.

MAP 25. SPANISH LAND GRANTS

Hempstead, Fay. *Historical Review of Arkansas: Its Commerce, History and Modern Affairs.* 1:22-27. Chicago: Lewis Publishing Company, 1911.

Herndon, Dallas T., ed. *Centennial History of Arkansas.* Vol. 1, chap. 7. Chicago and Little Rock: S. J. Clarke Publishing Company, 1922.

Rankin, Claude A., "Spanish Land Grants in Arkansas," *Pulaski County Historical Review* 13 (June, 1965): 24.

MAP 26. TOWNSHIP AND RANGE SYSTEM

Higgins, J. S. *Subdivisions of the Public Land.* Saint Louis, Mo.: Higgins and Company, 1894.

Logan, Robert R. "Notes on the First Land Survey in Arkansas." *Arkansas Historical Quarterly* 20 (Autumn, 1960): 260-71.

MAPS 27 AND 28. TERRITORIAL BOUNDARY CHANGES, 1803-1819 AND 1819-1836

Baird, W. David. "Arkansas' Choctaw Boundary: A Study of Justice Delayed." *Arkansas Historical Quarterly* 28 (Autumn, 1969): 203-22.

Herndon, Dallas T., ed. *Centennial History of Arkansas.* 1:11, 135-49, 162-95. Chicago and Little Rock: S. J. Clarke Publishing Company, 1922.

MAPS 29, 30, AND 31. ARKANSAS COUNTY FORMATION

A Digest of the Statutes of Arkansas Embracing All Laws of a General and Permanent Character in Force at the Close of the Session of the General Assembly of One thousand Eight Hundred and Eighty-three. Little Rock: Mitchell & Bettis, 1884.

Steele, J., and J. M'Campbell, comps. *Laws of Arkansas Territory.* Little Rock, Ark. Terr.: J. Steele, Esq., 1835.

MAP 32. INITIAL SETTLEMENT PATTERNS IN ARKANSAS TERRITORY

Hempstead, Fay. *Historical Review of Arkansas: Its Commerce, Industry, and Modern Affairs.* 1:49. Chicago: Lewis Publishing Company, 1911.

Morris, Wayne. "Traders and Factories on the Arkansas Frontier, 1805-1822." *Arkansas Historical Quarterly* 28 (Spring, 1969): 28-48

MAPS 33 AND 34. EARLY TRAILS AND MILITARY ROADS AND POST ROADS IN FRONTIER ARKANSAS

Herndon, Dallas T., ed. *Centennial History of Arkansas.* 1:505-19. Chicago and Little Rock: S. J. Clarke Publishing Company, 1922.

Longnecker, Julie Ward. "A Road Divided: From Memphis to Little Rock Through the Great Mississippi Swamp." *Arkansas Historical Quarterly* 34 (Autumn, 1985): 203-19.

Mitchell, S. Augustus. "A New Map of Arkansas with its Canals, Roads, and Distances." (Philidelphia). Date Unknown.

Moffatt, Walter. "Transportation in Arkansas, 1819–1840." *Arkansas Historical Quarterly* 40 (Autumn, 1956): 187–201.
Yarborough, Ann Nash. "Arkansas' Struggle for Communication." *Arkansas Historical Quarterly* 18 (Spring, 1959): 44–49.

MAP 35. NAVIGABLE WATERWAYS, 1860

Moffatt, Walter. "Transportation in Arkansas, 1819–1840." *Arkansas Historical Quarterly* 15 (Autumn, 1956): 187–210.

MAP 36. ANTEBELLUM AGRICULTURE

United States Bureau of the Census. *The Seventh Census of the United States, 1850.* Table 11. Washington, D.C.: Robert Armstrong, Public Printer, 1853.
______. *Agriculture of the United States in 1860; Compiled from the Original Returns of the Eighth Census.* Pp. 6–9. Washington, D.C.: United States Government Printing Office, 1864.

MAP 37. MIGRATION INTO ARKANSAS, 1830–1860

Pitcaithley, Dwight. "Settlement of the Arkansas Ozarks; the Buffalo River Valley." *Arkansas Historical Quarterly* 37 (Autumn, 1970): 203–20.
Walz, Robert. "Arkansas Slaveholdings and Slaveholders in 1850." *Arkansas Historical Quarterly* 12 (Spring, 1953): 38–73.
______. "Migration into Arkansas, 1834–1880." Ph.D. diss., University of Texas at Austin, 1958.
______. "Migration into Arkansas, 1820–1880: Incentives and Means of Travel." *Arkansas Historical Quarterly* 17 (Winter, 1959): 309–24.

MAP 38. SETTLEMENT IN FRONTIER ARKANSAS, 1830 -1860

Paullin, Charles. *Atlas of the Historical Geography of the United States.* Washington, D.C.: Carnegie Institution and American Geographical Society, 1932.

MAP 39. SECTIONAL POLITICS IN THE CONSTITUTIONAL CONVENTION OF 1836

Ledbetter, Cal, Jr. "The Constitution of 1836: A New Perspective." *Arkansas Historical Quarterly* 41 (Autumn, 1982): 215–52.
White, Lonnie J. *Politics on the Southwestern Frontier: Arkansas Territory, 1819–1836.* pp. 183–200. Memphis, Tenn.: Memphis State University Press, 1964.

MAP 40. DEMOCRATS AND WHIGS IN THE 1844 GUBERNATORIAL CAMPAIGN

Boyett, Gene W. "The Whigs of Arkansas, 1836–1856. "Ph.D. diss. Louisiana State University, 1972.
______. "Quantitative Differences Between the Arkansas Whig and Democratic Parties, 1836–1850." *Arkansas Historical Quarterly* 34 (Autumn, 1975): 214–26.
Meek, Melinda. "The Life of Archibald Yell." *Arkansas Historical Quarterly* 26 (Winter, 1967): 353–78.
Stokes, D. A. "Public Affairs in Arkansas, 1836–1850," Ph.D. diss. University of Texas at Austin, 1966.
Walton, Brian G. "The Second Party System in Arkansas, 1836–1848," *Arkansas Historical Quarterly* 33 (Summer, 1969): 120–55.

MAP 41. SECESSIONIST POLITICAL PATTERNS

"Delegates Disclose Their Allegiance," *Arkansas Gazette,* February 20–22, 1861.
Scroggs, Jack. "Arkansas in the Secession Crisis." *Arkansas Historical Quarterly* 12 (Autumn, 1953): 179–224.
Wooster, Ralph A.. "The Arkansas Secession Convention." *Arkansas Historical Quarterly* 13 (Autumn 1954): 172–95.

MAP 42. MAJOR CIVIL WAR BATTLES

Dougan, Michael B. *Confederate Arkansas, The People and Policies of a Frontier State in Wartime.* University, Ala.: University of Alabama Press, 1976.
Ferguson, John L., ed. *Arkansas and the Civil War.* Little Rock: Pioneer Press, 1962.
Harrell, John M. *Confederate Military History.* Vol. 10. Atlanta, Ga.: Confederate Publishing Company, 1899.
Thomas, David Y. *Arkansas in War and Reconstruction, 1861–1874.* Little Rock: Arkansas Division, United Daughters of the Confederacy, 1926.

MAP 43. CIVIL WAR BATTLESITES: PEA RIDGE

Bearss, Edwin. "The First Day at Pea Ridge, March 7, 1862. *Arkansas Historical Quarterly* 17 (Spring, 1958): 132–54.
Brown, Walter Lee. "Pea Ridge: Gettysburg of the West." *Arkansas Historical Quarterly* 15 (Spring, 1956): 3–16.

MAP 44. CIVIL WAR BATTLESITES: PRAIRIE GROVE

Monnett, Howard N. "A Yankee Cavalryman Views the Battle of Prairie Grove." *Arkansas Historical Quarterly* 21 (Winter, 1962): 289–304.
Oates, Stephen. "The Prairie Grove Campaign, 1862." *Arkansas Historical Quarterly* 19 (Summer, 1960): 119–41.

MAP 45. CIVIL WAR BATTLESITES: ARKANSAS POST

Bearss, Edwin C., "The Battle of the Post of Arkansas." *Arkansas Historical Quarterly* 18 (Autumn, 1959): 237–79.

MAP 46. CIVIL WAR BATTLESITES: HELENA

Bearss, Edwin. "The Battle of Helena, July 4, 1863." *Arkansas Historical Quarterly* 20 (Autumn, 1961): 256–297.

MAP 47. CIVIL WAR BATTLESITES: LITTLE ROCK

Huff, Leo. "The Union Expedition Against Little Rock, August–September, 1863." *Arkansas Historical Quarterly* 22 (Autumn, 1963): 224–37.

MAP 48. CIVIL WAR BATTLESITES: JENKINS' FERRY

Richards, Ira Don. "The Battle of Jenkins' Ferry." *Arkansas Historical Quarterly* 20 (Spring, 1961): 3–16.
______. "The Battle of Poison Springs." *Arkansas Historical Quarterly* 18 (Winter, 1959): 338–349.
______. "The Engagement at Marks' Mill." *Arkansas Historical Quarterly* 19 (Spring, 1960): 51–60.

MAP 49. MAJOR RAILROADS, 1915

Arkansas Railroad Commission. *First Annual Report of the Railroad Commission of the State of Arkansas.* Little Rock: Thompson Lithograph and Printing Co., 1901.
______. *Report of the Railroad Commission, 1931–32.* Texarkana: Helms Printing Co., 1933.
Dew, Lee A., and Louis Koeppe. "Narrow-Gauge Railroads in Arkansas." *Arkansas Historical Quarterly* 21 (Autumn, 1972): 276–93.
Herndon, Dallas T., ed. *Centennial History of Arkansas.* 1:519–27. Chicago and Little Rock: S. J. Clarke Publishing Company, 1922.
Wood, Stephen E. "The Development of Arkansas Railroads." Parts 1, 2. Arkansas Historical Quarterly 75 (Summer, Autumn, 1948): 103–40, 155–93.

MAP 50. SHORT-LINE RAILROADS, 1876–1914

Hull, Clifton. *Shortline Railroads of Arkansas.* Norman: University of Oklahoma Press, 1969.

MAP 51. TIMBER INDUSTRY

United States Bureau of the Census. *Reports on Manufactures: 1940.* Vol. 3, *Reports of the States,* table 9, p. 76. Washington, D.C.: United States Government Printing Office, 1942.

MAPS 52 AND 53. BAUXITE AND FOSSIL-FUEL MINING, AND OTHER MINERAL RESOURCES

Arkansas Geological Survey. *Annual Report of the Geological Survey of Arkansas,* various volumes.

Pitcaithley, Dwight, "Zinc and Lead Mining Along the Buffalo River." *Arkansas Historical Quarterly* 37 (Winter, 1978): 293–305

United States Department of the Interior. *Mineral Resources Year Book: 1984,* 87–95. Washington, D.C.: United States Government Printing Office, 1986. See also previous years.

MAP 54. NEW SOUTH AGRICULTURAL PATTERNS

United States Bureau of the Census. *Report on Cotton Production in the United States; Also Embracing Agriculture and Physio-Geographical Descriptions of the Several Cotton States and of California.* Part 1, *Mississippi Valley and Southwestern States.* Washington D.C.: United States Government Printing Office, 1884.

______. *Twelfth Census of the United States, Taken in the Year 1900. Agriculture.* Part 2, *Crops and Irrigation,* Tables 10, 15, 22, 40, 55. Washington, D.C.: United States Government Printing Office, 1902.

MAP 55. AGRICULTURAL PATTERNS, 1900–1949

United States Bureau of the Census. *Abstract of the Fifteenth Census of the United States.* tables 86–144. Washington D.C.: United States Government Printing Office, 1933.

______. *United States Census Agriculture: 1950. General Report.* Vol. 3. Washington D.C.: United States Government Printing Office, 1952.

______. *United States Cenus of Agriculture: 1950. Counties and State Economic Areas.* Vol. 1, pt. 23, table 11. Washington, D.C.: United States Government Printing Office, 1952.

MAP 56. FARM TENANCY, 1910

Holley, Donald. *Uncle Sam's Farmers: The New Deal Communities in the Lower Mississippi Valley.* Chicago: University of Illinois Press, 1975.

Norse, E. G. "A Preliminary Survey of Land Tenure in Arkansas." University of Arkansas Extension Service Bulletin. Fayetteville, 1919.

United States Bureau of the Census. *United States Census of Agriculture: 1950. General Report,* Vol. 3. Washington, D.C.: United States Government Printing Office, 1952.

______. 1978 Census of Agriculture. Vol. 1, *State and County Data,* pt. 4, "Arkansas." Washington, D.C.: United States Government Printing Office, 1979.

MAP 57. MIGRATION INTO ARKANSAS, 1860–1880

Walz, Robert. "Migration into Arkansas, 1834–1880." Ph.D. diss., University of Texas at Austin, 1958.

MAP 58. EUROPEAN ETHNIC SETTLEMENTS

Felton, Freeman D. "Immigration to Arkansas." *Arkansas Historical Quarterly* 7 (Autumn, 1948), 210–20.

Herndon, Dallas T., ed. *Centennial History.* 1:397–98. Chicago and Little Rock: S. J. Clarke Publishing Company, 1922.

Lensing, Michael. "The Founding of Subiaco Abbey." *Arkansas Historical Quarterly* 3 (Autumn, 1944): 193–210.

Lewellen, Jeffrey. " 'Sheep Amidst the Wolves': Father Bandini and the Colony at Tontitown, 1898–1917." *Arkansas Historical Quarterly* 45 (Spring, 1986): 19–40.

Rothrock, Thomas. "The Founding of Tontitown, Arkansas." *Arkansas Historical Quarterly* 16 (Spring, 1957): 20–35.

Sarna, Jean. "Marche, Arkansas: A Personal Reminiscence of Life and Customs." *Arkansas Historical Quarterly* 36 (Spring, 1977): 31–49.

Thomas, David. "The Chinese Labor Question, 1869–1870." *Pulaski County Historical Review* 30 (1982): 59–63.

Wolfe, Jonathan James. "Background of German Immigration." Parts 1–3. *Arkansas Historical Quarterly* 25 (Spring, Summer, Autumn, 1966): 151–82, 248–78, 354–86.

MAP 59. BLACK POPULATION SINCE THE CIVIL WAR

United States Bureau of the Census. *Negro Population, 1790–1915.* Washington, D.C. United States Government Printing Office, 1918.

MAP 60. GENERAL POPULATION TRENDS, 1860–1950

United States Bureau of the Census. *Historical Statistics of the United States, Colonial Times to 1970, Bicentennial Edition.* p. 24. Washington, D.C.: United States Government Printing Office, 1975. See specific decenniel census reports for county details.

MAP 61. FARM PROTEST: THE GUBERNATORIAL ELECTION OF 1888

Davis, Granville D. "The Granger Movement in Arkansas." *Arkansas Historical Quarterly* 4 (Winter, 1945): 340–52.

Paisley, Clifton. "The Political Wheelers and Arkansas' Election of 1888." *Arkansas Historical Quarterly* 25 (Spring, 1966): 3–21.

MAP 62. HIGHWAY CONSTRUCTION, 1925–1955

Arkansas Legislative Council, Committee on Roads and Highways. *The Public Roads of Arkansas and Their Use.* 1957.

Arkansas State Highway Commission. *Thirty-Fifth Biennial Report: Fiscal Years, July 1, 1980, Through June 30, 1981, July 1, 1981 through June 30, 1982.*

MAP 63. AGRICULTURAL REGIONS

Atlas of Arkansas. P. 24. Little Rock: Arkansas Department of Planning, 1973.

Dethloff, Henry C. "Rice Revolution in the Southwest, 1880–1910." *Arkansas Historical Quarterly* 29 (Spring, 1970), 66–75.

Fielder, V. B. *Types of Farming Areas in Arkansas.* Bulletin 555. Fayetteville: University of Arkansas Experiment Station, 1955.

MAP 64. ROW CROPS SINCE 1950

United States Bureau of the Census. *United States Census of Agriculture: 1950. General Report.* Vol. 3. Washington, D.C.: United States Government Printing Office, 1952.

______. *1974 Census of Agriculture.* Vol. 1, pt. 4. *Arkansas State and County Data.* Washington, D.C.: United States Government Printing Office, 1974.

MAP 65. EXPANSION OF MANUFACTURING: THE ELECTRICAL AND ELECTRONIC EQUIPMENT INDUSTRIES

United States Bureau of the Census. *Census of Manufactures, 1966.* Vol. 3, *Area Statistics.* Washington, D.C.: United States Government Printing Office, 1966.

______. *1977 Census of Manufactures.* Washington, D.C.: United States Government Printing Office, 1977.

MAP 66. GENERAL POPULATION GROWTH SINCE 1950

United States Bureau of the Census. *Historical Statistics of the United States, Colonial Times to 1970, Bicentennial Edition.* P. 24. Washington, D.C.: United States Government Printing Office, 1975. See specific decenniel census reports for county details.

______. *1980 Census of Population. Characteristics of the Population. General Social and Economic Characteristics.* Vol. 1, chap. 3, pt. 5, "Arkansas." Washington, D.C.: United States Government Printing Office, 1983.

MAP 67. URBANIZATION, 1900–1980

United States Bureau of the Census. *Historical Statistics of the United States, Colonial Times to 1970, Bicentennial Edition.* Pts. 1, 24. Washington, D.C.: United States Government Printing Office, 1975.

______. *1980 Census of Population. Characteristics of the Population. General Social and Economic Characteristics.* Vol. 1, chap. 3, pt. 5, "Arkansas." Washington, D.C.: United States Government Printing Office, 1983.

MAP 68. THE 1966 GUBERNATORIAL ELECTION

Jones, Merrill Andy. "A Rhetorical Study of Winthrop Rockefeller's Political Speeches, 1964 through 1971." Ph.D. diss., Louisiana State University, 1984.

Ward, John. *The Arkansas Rockefeller.* Baton Rouge: Louisiana State University Press, 1978.

MAPS 69, 70, AND 71. THE 1968, 1972, AND 1980 PRESIDENTIAL RACES

Alber, Connie. *The Wallace Movement in Arkansas.* Little Rock, Ark.: Southern Political Research Bureau, 1970.

Johnston, Robert. "The 1980 Election in Arkansas." Paper presented to the Arkansas Political Science Association, 1981.

Martin, Ward M. "Arkansas's Aging Trend: A Portrait of Future Politics." *Southwestern Social Sciences Quarterly* 45 (June, 1964): 37–49.

Savage, Robert L., and Diane D. Blair. "Regionalism and Political Opinion in Arkansas: An Exploratory Survey." *Arkansas Political Science Journal* 5 (Winter, 1984): 59–85.

Yates, R. "Arkansas: Independant and Unpredictable." *In The Changing Politics of the South,* edited by W. C. Havard. Baton Rouge: Louisiana State University Press, 1972.

INDEX

The numbers refer to maps. Italic numbers indicate placement on a map.